农业英语

（第2版）

主　编　王静萱
副主编　张慧琴　陈延森

重庆大学出版社

内容简介

《农业英语》是"甘肃省教育科学'十二五'规划课题"研究成果。本教材为高等学校农业类专业学生学习专业英语编写,农、林、生物等专业的学生都可选用。同时,也可供各类成人院校及广大农林企业从业人员学习专业英语,提高涉外业务交际能力使用。

全书由10个情境教学组成,内容选择上难易结合,循序渐进,涉及农业、植物、植物生理、食品、农业技术、现代农业、种植、植物保护、生态农业以及应用写作等方面的知识。每个情境教学由3~5篇文章组成,题材实用,相关知识都配有图片,帮助学生记忆和理解。为便于学习者学习,每篇文章均配有难句和要点注释、练习、参考译文及参考答案,以提高学生专业英语综合能力,帮助专业人员自学。附录部分增加了作物和蔬菜水果、农业经济等专业术语,便于学习者查阅和应用。

图书在版编目(CIP)数据

农业英语/王静萱主编.--2版.--重庆:重庆大学出版社,2021.1

ISBN 978-7-5624-8933-7

Ⅰ.①农… Ⅱ.①王… Ⅲ.①农业—英语—高等职业教育—教材 Ⅳ.①S

中国版本图书馆CIP数据核字(2021)第016131号

农业英语
(第2版)
主 编 王静萱
策划编辑:杨粮菊
责任编辑:范 琪 荀荟羽 版式设计:杨粮菊
责任校对:关德强 责任印制:张 策
*
重庆大学出版社出版发行
出版人:饶帮华
社址:重庆市沙坪坝区大学城西路21号
邮编:401331
电话:(023) 88617190 88617185(中小学)
传真:(023) 88617186 88617166
网址:http://www.cqup.com.cn
邮箱:fxk@cqup.com.cn (营销中心)
全国新华书店经销
重庆俊蒲印务有限公司印刷
*
开本:787mm×1092mm 1/16 印张:14.75 字数:405千
2015年3月第1版 2021年1月第2版 2021年1月第4次印刷
ISBN 978-7-5624-8933-7 定价:49.80元

本书如有印刷、装订等质量问题,本社负责调换
版权所有,请勿擅自翻印和用本书
制作各类出版物及配套用书,违者必究

前　言

《农业英语》是“甘肃省教育科学‘十二五’规划课题”研究成果。

为实现高职教育应用型人才培养的总体目标，满足行业企业对农业技术类人才专业英语知识的需求，贯彻《教育部关于以就业为导向深化高等职业教育改革的若干意见》精神，特编写此书。本书力求将区域农业外向型发展特色与学生未来工作岗位实际有机结合，旨在为学生提供岗位所需的专业英语知识和英语技能，介绍农业领域的新信息，激发学生基于已有知识获取新知识的创新学习能力，以此培养他们专业英语综合能力和涉外业务处理能力。

《农业英语》为农科院校高等职业学生学习英语编写，农、林、生物、作物种植、农产品加工等专业的学生都可选用。鉴于本书注重英语综合能力和交际技能的特点，也可用于农林企业从业人员专业培训。

本书在农业技术类人才专门用途英语(ESP)教学研究与实践的基础上编写，具有较强的实用性和针对性。注重农业基础知识的同时紧跟当前形势，适度介绍农业前沿信息。

内容选择上难易结合，循序渐进，全书共10个单元，涉及农业、植物、植物生理、食品、农业技术、现代农业、种植、植物保护、生态农业以及应用写作等方面。每单元由3~5篇文章组成，题材实用，均配有相关注释与练习，以提高学生专业英语综合能力，帮助专业人员自学。增加了农业经济、作物和蔬菜水果名称，便于学习者查阅。

为改变专门用途英语教学沉闷枯燥无效的现状，本书每个单元都配有情景会话可以在单元开始前作为热身练习；课后练习旨在提升农业英语综合能力，既围绕情景展开，又有适当拓展，教学中教师可视具体情况灵活选用。

本书由王静萱担任主编，收集与整理了所有资料，编写了Unit 1、Unit 2、Unit 3、Unit 6、Unit 8及附录的部分内容。张慧琴担任副主编，编写了Unit 7和Unit 9。陈延森担任副主编，编写了Unit 4、Unit 5、Unit 10及附录的部分内容，同时负责全书的校对工作。

在编著本书的过程中，编者参考了大量国内外有关书籍和资料，特别是网络资料，在此表示衷心的感谢。由于水平有限，疏漏之处敬请专家读者批评指正。

编者

2020年9月

Contents

Unit 1

Agriculture

> *Agriculture is the lifeline of the national economy, also is the national industrialization base.*
>
> ——农业是国民经济的命脉,也是国家实现工业化的基础。

Situational Dialogue:

Visiting a Modern Household Farm

A: Good morning, Mrs. Brown. Welcome to our farm.

B: Good morning, Mr. Chen. I thought you said a couple of acres, a vegetable patch and a few chickens. But it is a big farm.

A: Yes, we have a two-story house and almost a hundred acres under wheat, and also a cow barn. It's lighted and heated and as clean as an operating theater.

B: This is not just a farm. It's a big business.

A: Yes. All the cultivation is mechanized. The harvesting machines are twice the size of your living room.

B: I guess you raise pigs and chickens, too?

A: That's right. And we sell the eggs and we also keep our own bees. We take care of the bees ourselves and process the honey.

A: You sound like a remarkable farmer.

B: Thank you. I like the farm and the farm life.

A: Thanks a lot. Good-bye.

B: Bye.

a vegetable patch:一块菜地

a cow barn：牛舍

Lesson 1 Agriculture

Agriculture is the **process** of producing food, including **grains**, **fiber**, fruits, and **vegetables**, as well as feed for animals. It also includes **raising livestock—domesticated** animals such as cows. Besides food for humans and animal feeds, agriculture produces goods, such as flowers, **nursery** plants, **timber**, **leather**, **fertilizers**, fibers (such as **cotton** and **wool**), **fuels** (such as **biodiesel**), and **drugs** (such as **aspirin**, **sulfa**, and **penicillin**).

Agriculture provides food, feed, fiber, fuel, furniture, raw materials and materials for and from factories; provides a free **fresh environment**, **abundant** food for driving out **famine**; **favors friendship** by **eliminating fights**. Satisfactory agricultural production brings peace, prosperity, harmony, health and wealth to individuals of a nation by driving away **distrust**, **discord** and **anarchy**.

Agriculture helps to meet the basic needs of human beings and their **civilization** by providing food, clothing, shelters, medicine and **recreation**. Hence, agriculture is the most important **enterprise** in the world. It is a productive unit where the free gifts of nature namely land, light, air, temperature and rain water, etc. are integrated into single primary unit indispensable for human beings. Secondary productive units namely animals including livestock, birds and insects, feed on these primary units and provide products such as meat, milk, wool, eggs, honey, silk and lac.

1. Terminology

Agriculture is **derived from** Latin words Ager and Cultura. Ager means land or field and Cultura means cultivation. Therefore the term agriculture means cultivation of land, i.e., the science and art of producing crops and livestock for economic purposes. It is also referred as the science of producing crops and livestock from the natural resources of the earth. The primary aim of agriculture is to cause the land to produce more abundantly, and at the same time, to protect it from **deterioration** and **misuse**. It is **synonymous** with farming—the production of food, **fodder** and other industrial materials.

2. Branches of Agriculture

Crop Production—It deals with the production of various crops, which includes food crops, fodder crops, fibre crops, sugar, oil seeds, etc. It includes **agronomy**, soil science, **entomology**, **pathology**, **microbiology**, etc. The aim is to have better food production and to control the diseases.

Horticulture—A branch of agriculture deals with the production of flowers, fruits, vegetables, **ornamental** plants, **spices**, condiments and **beverages**.

Agricultural Engineering—It is an important component for crop production and horticulture particularly to provide tools and **implements**. It is aiming to produce modified tools to facilitate proper animal husbandry and crop production.

Forestry—It deals with production of large scale cultivation of **perennial trees** for supplying wood, timber, rubber, etc. and also raw materials for industries.

Animal Husbandry—It is the science of looking after and breeding animals—specifically those that are used in agriculture, to provide products, for research purposes or as domestic pets.

Fishery Science—It is for marine fish and inland fishes including **shrimps** and **prawns**.

Vocabulary

process [ˈprəʊses] *n.* 过程
grain [greɪn] *n.* 谷物,粮食
fiber [ˈfaɪbə] *n.* 纤维,纤维物质
vegetable [ˈvedʒtəbl] *n.* 蔬菜
raise [reɪz] *v.* 饲养
livestock [ˈlaɪvstɒk] *n.* 家畜,牲畜
domesticate [dəˈmestɪkeɪt] *v.* 驯养

nursery [ˈnɜːsəri] *n.* 苗圃
timber [ˈtɪmbə] *n.* 木材,木料
leather [ˈleðə] *n.* 皮革
fertilizer [ˈfɜːtəlaɪzə] *n.* 肥料
cotton [ˈkɒtn] *n.* 棉花
wool [wʊl] *n.* 羊毛
fuel [fjʊəl] *n.* 燃料
biodiesel [ˈbaɪəʊdiːzl] *n.* 生物柴油
drug [drʌg] *n.* 药物
aspirin [ˈæsprɪn] *n.* 阿司匹林
sulfa [ˈsʌlfə] *n.* [药]磺胺制剂
penicillin [ˌpenɪˈsɪlɪn] *n.* 青霉素,盘尼西林
fresh [freʃ] *adj.* 新鲜的；淡水的
environment [ɪnˈvaɪrənmənt] *n.* 环境
abundant [əˈbʌndənt] *adj.* 大量的；丰富的
famine [ˈfæmɪn] *n.* 饥荒；饥饿
favor [ˈfeɪvə] *v.* 偏爱；赞同
friendship [ˈfrendʃɪp] *n.* 友情,友谊
eliminate [ɪˈlɪmɪneɪt] *v.* 消除
fight [faɪt] *n.* 战斗
distrust [dɪsˈtrʌst] *n.* 不信任,猜疑
discord [ˈdɪskɔːd] *n.* 不和
anarchy [ˈænəki] *n.* 混乱
civilization [ˌsɪvəlaɪˈzeɪʃn] *n.* 文明
recreation [ˌrekriˈeɪʃn] *n.* 消遣；娱乐
enterprise [ˈentəpraɪz] *n.* 事业
terminology [ˌtɜːmɪˈnɒlədʒi] *n.* 术语
derive from 起源于
deterioration [dɪˌtɪərɪəˈreɪʃn] *n.* 恶化
misuse [ˌmɪsˈjuːz] *n.* 滥用
synonymous [sɪˈnɒnɪməs] *adj.* 同义的
fodder [ˈfɒdə] *n.* 草料
agronomy [əˈgrɒnəmɪ] *n.* 农艺学；农学
entomology [ˌentəˈmɒlədʒi] *n.* 昆虫学
pathology [pəˈθɒlədʒi] *n.* 病理(学)
microbiology [ˌmaɪkrəʊbaɪˈɒlədʒi] *n.* 微生物学
horticulture [ˈhɔːtɪkʌltʃə] *n.* 园艺(学)

ornamental [ˌɔːnəˈmentl] *adj.* 装饰的
spice [spaɪs] *n.* 香料
condiment [ˈkɒndɪmənt] *n.* 调味品
beverage [ˈbevərɪdʒ] *n.* 饮料
implement [ˈɪmplɪment] *n.* 工具
perennial tree 多年生树木
shrimp [ʃrɪmp] *n.* 虾
prawn [prɔːn] *n.* 对虾,明虾

Notes

1. domesticated animals 驯养的动物,家养的动物
2. Agriculture helps to meet the basic needs of human beings and their civilization by providing food, clothing, shelters, medicine and recreation.
   农业通过提供食物、衣服、住所、药品和娱乐,有助于满足人类和人类文明的基本需求。
3. Agriculture is derived from Latin words Ager and Cultura.
   Agriculture 源于拉丁语的单词 Ager 和 Cultura。
   be derived from 源于

Exercises

Ⅰ. Review the text and translate the following words and phrases into Chinese.

1. grain
2. fiber
3. fruit
4. vegetable
5. fodder
6. nursery plant
7. timber
8. leather
9. fertilizer
10. prosperity
11. wool
12. fuel
13. biodiesel
14. drug
15. aspirin
16. sulfa
17. penicillin
18. raw material
19. peace
20. harmony
21. health
22. wealth

Ⅱ. Match the following words and phrases.

1. agriculture　　A. 作物生产
2. fodder　　B. 园艺

3. cotton	C. 林业
4. crop production	D. 农业
5. horticulture	E. 蔬菜
6. forestry	F. 棉花
7. animal husbandry	G. 草料
8. fishery science	H. 水果,果实
9. fruit	I. 畜牧业
10. vegetable	J. 渔业科学

Ⅲ. Fill in the missing words.

Agriculture helps to 1. ________ the basic 2. ________ of human and their civilization by 3. ________ food, clothing, shelters, medicine and recreation. Hence, agriculture is the most important 4. ________ in the world. It is a productive unit where the 5. ________ gifts of nature namely 6. ________, light, air, temperature and rain water etc., are 7. ________ into single primary unit indispensable for human beings. Secondary productive units namely 8. ________ including livestock, birds and 9. ________, feed on these primary units and provide concentrated products 10. ________ meat, milk, wool, eggs, honey, silk and lac.

Ⅳ. Read the following passage and then answer the questions.

The word agronomy has been derived from the two Greek words, agros and nomoshaving, the meaning of field and managing, respectively. Literally, agronomy means the "art of managing field". Technically, it means the "science and economics of crop production by management of farm land".

A. Definition

Agronomy is the art and underlying science in production and improvement of field crops with the efficient use of soil fertility, water, labourer and other factors related to crop production.

Agronomy is the field of study and practice of ways and means of production of food, feed and fiber crops. Agronomy is defined as "a branch of agricultural science which deals with principles and practices of field crop production and management of soil for higher productivity".

B. Importance

Among all the branches of agriculture, agronomy occupies a pivotal position and is regarded as the mother branch or primary branch. Like agriculture, agronomy is an integrated and applied aspect of different disciplines of pure sciences. Agronomy has three clear branches namely, (ⅰ) crop science, (ⅱ) soil science, and (ⅲ) environmental science that deals only with applied aspects, (i. e.) soil-crop-environmental relationship. Agronomy is a synthesis of several disciplines like crop science, which includes plant breeding, crop

physiology and biochemistry, etc., and soil science, which includes soil fertilizers, manures, etc., and environmental science which includes meteorology and crop ecology.

Questions:

1. How do you understand the word "agronomy"?

2. What is agronomy?

3. Why is agronomy important?

V. Please use your knowledge to decribe your major. Don't forget to use as much as what you've learned from the text.

参考译文

农 业

农业是生产食品包括谷物、纤维、水果、蔬菜以及动物饲料的过程。它还包括饲养牲畜——驯养的牛等动物。除了人类食品和动物饲料之外,农业还生产如花卉、苗木、木材、皮革、肥料、纤维(如棉、羊毛)、燃料(如生物柴油)以及药物(如阿司匹林、磺胺类、青霉素)。

农业提供食品、饲料、纤维、燃料、家具、原材料和工厂所需的材料;提供免费清新的环境,以及丰富的食物避免饥荒;通过消除战争加强友谊。良好的农业生产为一国居民带来和平、繁荣、和谐、健康和财富,赶走猜疑、不和谐与混乱。

农业通过提供食物、衣服、住所、医学和娱乐,有助于满足人类和人类文明的基本需求。因此,农业是世界上最重要的产业。这是一个多产的产业,所有免费的自然馈赠如土地、光、空气、温度和雨水等,都融合到人类不可缺少的基础产业(种植业)中。次级多产单元即动物,包括家畜、鸟类和昆虫,以这些基础单元为食,以及供应如肉类、牛奶、羊毛、鸡蛋、蜂蜜、丝绸和紫胶等制成品。

1. 术语

农业是来自拉丁语的单词 Ager 和 Cultura。Ager 是指土地,Cultura 意思是培养。因此,农业这一术语意思是土地的耕种和培养,也就是出于经济的目的生产作物和牲畜的科学和艺术。

它也是指从地球的自然资源中生产作物和牲畜的科学。农业的主要目的是使土地更多产,同时防止土地恶化和滥用。它是 farming 的同义词——生产食品、饲料和其他工业原材料。

2. 农业的分支

作物生产——进行各种作物的生产,包括粮食作物、饲料作物、纤维作物、糖、油种子等。它包括农学、土壤学、昆虫学、病理学、微生物学等。其目的是为了有更好的食品生产以及控制疾病。

园艺——主要是进行花卉、水果、蔬菜、花卉、香料、调味品(包括有药用价值的鸦片毒品作物等)和饮料生产的农业分支。

农业工程——是作物生产和园艺的一个重要组成部分,可以为农业提供工具。它的目标是生产改良的工具来促进动物饲料和作物生产。

林业——主要是为供应木材、橡胶和工业原料而进行多年生树木的大规模生产。

畜牧业——是照料和饲养动物的科学,特别是那些在农业中用到的出于研究的目的或作为家庭宠物的动物饲养,并提供相关产品。

渔业科学——主要是海洋鱼类和内陆鱼虾。

Lesson 2 How Agriculture Has Changed Civilization

When people were living the **hunter-gatherer lifestyle**, the land could only support a limited number of people. Once crops could be grown, **harvested**, and **stored**, all that changed. The **advantages** of beginning an agricultural society were that a larger **population** could be supported and the chances of **survival** were **enhanced** by being able to store **excess** food over the winter. Agriculture also allowed people to stay in one place and not have to move around to gather food. It **promoted commerce** (business and trading) between civilizations, which were able to sell goods and make money. This began the early stages of modernizing the world.

All the major centers of agriculture began along major river systems. Without rivers like the Nile, the Indus, the Huang, the Tigris, and the Euphrates to provide a **consistent** source of **silt** (a natural fertilizer) from yearly **floods**, and water for **irrigating** crops, agricultural development could not have taken place. It was also during this time that our ancestors realized they could grow rice on flooded fields.

As farming became more **sophisticated**, fewer people needed to be farmers. This freed others to **pursue** scientific, industrial, and cultural paths, which led to many new inventions. This shift made possible developments in **architecture**, including the building of the huge palaces, temples, and theaters for which many famous ancient sites are known.

Take, for example, the majestic pyramids of Egypt and the beautiful temples in Greece. Advances in agriculture allowed other people to become scientists and study **astronomy**, which began the development of **navigational** skills that were later used to **explore** the world. None of the major human developments through history would have been possible if agriculture had not been developed. Farming led to a food **surplus** that could support artists, builders, philosophers, and scientists.

Later, the Industrial Revolution in the late eighteenth century caused the rapid growth of towns and cities and forced agriculture to be isolated within its own area. As inventions like the cultivator, reaper, thresher, and combine appeared, modern agriculture further advanced. These advances enabled large-scale agriculture to develop. Modern science also **revolutionized** food processing, such as with the invention of **refrigeration**. Today, harvesting operations have been mechanized for almost every plant product. Breeding programs have made livestock production more efficient, too. Genetic engineering has revolutionized growing crops and raising livestock. Agriculture has played a significant role in allowing people to have the lifestyles and freedoms they enjoy today.

Vocabulary

hunter-gatherer [ˈhʌntə ˈɡæðərə] *n.* 依靠狩猎和采集生活的人
lifestyle [ˈlaɪfstaɪl] *n.* 生活方式
harvest [ˈhɑːvɪst] *v.* 收割
store [stɔː] *v.* 储存
advantage [ədˈvɑːntɪdʒ] *n.* 益处；优越(性)
population [ˌpɒpjuˈleɪʃn] *n.* 人口
survival [səˈvaɪvl] *n.* 幸存
enhance [ɪnˈhɑːns] *v.* 提高，增加；加强
excess [ɪkˈses] *adj.* 过量的，额外的
promote [prəˈməʊt] *v.* 促进，推进
commerce [ˈkɒmɜːs] *n.* 商业；贸易
consistent [kənˈsɪstənt] *adj.* 一致的；连续的
silt [sɪlt] *n.* 淤泥
flood [flʌd] *n.* 洪水
irrigate [ˈɪrɪɡeɪt] *v.* 灌溉

sophisticated [sə'fɪstɪkeɪtɪd] *adj.* 复杂的
pursue [pə'sjuː] *v.* 继续；追求
architecture ['ɑːkɪtektʃə] *n.* 建筑学
astronomy [ə'strɒnəmi] *n.* 天文学
navigational [ˌnævɪ'geɪʃənl] *adj.* 航行的，航海的
explore [ɪk'splɔː] *v.* 探索(险)
surplus ['sɜːpləs] *n.* 剩余，盈余
revolutionize [ˌrevə'luːʃənaɪz] *v.* 彻底改变，发动革命
refrigeration [rɪˌfrɪdʒə'reɪʃn] *n.* 冷藏，制冷

Notes

1. When people were living the hunter-gatherer lifestyle, the land could only support a limited number of people.
   the hunter-gatherer lifestyle：狩猎采集的生活方式；a number of 一些，若干。
2. Without rivers like the Nile, the Indus, the Huang, the Tigris, and the Euphrates to provide a consistent source of silt (a natural fertilizer) from yearly floods, and water for irrigating crops, agricultural development could not have taken place.
   ①the Nile：尼罗河，位于非洲；②the Indus：印度河，位于印度西北部；③the Tigris：底格里斯河，位于西南亚，流经土耳其和伊拉克；④the Euphrates：幼发拉底河，位于非洲。
3. Take, for example, the majestic pyramids of Egypt and the beautiful temples in Greece.
   以宏伟的埃及金字塔和美丽的希腊寺庙为例。
   take… for example 以……为例
4. Farming led to a food surplus that could support artists, builders, priests, philosophers, and scientists.
   Farming led to a food surplus… 此部分中 lead to 表示“导致、引起”，农业使粮食产生盈余；that 引导定语从句，修饰先行词 a food surplus，support 在这里表示“养活”，粮食的盈余养活了艺术家、建筑工人、哲学家和科学家。

Exercises

Ⅰ. Fill in the blanks with the words given below. Change the forms where necessary.

support	limit	enhance	promote	explore
pursue	harvest	role	isolate	store

1. Agriculture has played a significant ________________ in allowing people to have the lifestyles and freedoms they enjoy today.

2. Use daily writing to ________________ your communication skills.

3. The film ______________ the relationship between artist and instrument.

4. Millions of people are threatened with starvation as a result of drought and poor ________.

5. This policy could ______________ the country from the other permanent members of the United Nations Security Council.

6. I have children to ____________, money to be earned, and a home to be maintained.

7. You don't have to sacrifice (牺牲) environmental protection to ____________ economic growth.

8. She ____________ the man who had stolen a woman's bag.

9. When people were living the hunter-gatherer lifestyle, the land could only support a _____ ________ number of people.

10. He's ____________ away nearly one ton of potatoes.

Ⅱ. Read the text and decide whether each of the following statements is True (T) or False (F).

1. When people were living the hunter-gatherer lifestyle, the land could support almost all of the people.

2. All the major centers of agriculture began along major river systems.

3. It was also during this time that our ancestors realized they could grow rice.

4. As farming became more sophisticated, more and more people needed to be farmers.

5. Later, the Industrial Revolution in the late twentieth century caused the rapid growth of towns and cities and forced agriculture to be isolated within its own area.

Ⅲ. Translate the following paragraph.

Urban forestry involves forestry activities introduced from wildness and countryside to populous cities where economy, culture, industry and business are aggregated. Most metropolitan cities are currently prosperous but noisy with deteriorating ecological environment. People living in such a crowded and narrow space suffer from a worsening physical quality. To develop urban forestry can beautify living places, purify air, reduce noise and adjust local climate, so that the living quality of urban people can be improved. Generally, urban forestry has provided a new approach to the urban environmental problems.

Ⅳ. Discuss the relation between agriculture and civilizaiton according to what you have learned. And think about what it is like between modern agriculture and society.

参考译文

农业如何改变了文明

当人们以狩猎采集的生活方式生活时,土地只能养活有限数量的人。一旦作物可以种植、收获、储存,一切都改变了。进入农业社会的优势是通过储存多余的食物过冬供养更多的人,增强了他们存活的机会。农业也允许人们待在一个地方而不必为了寻找食物而迁徙。它促进了不同文化之间的商品和货币流通,从而使商业(商贸)得以发展。这就是现代化的世界的早期阶段。

所有主要的农业中心都诞生于主要河流沿岸。没有尼罗河、印度河、黄河、底格里斯河以及幼发拉底河等河流每年洪水冲刷形成的淤泥和灌溉农作物的话,农业的发展是不可能产生的。也正是在这个时候,我们的祖先意识到他们可以在被洪水淹没的田野里种植水稻。

随着农业变得越来越复杂,需要的农民越来越少。这使得其他人走上追求科学、工业和文化的道路,随之产生了许多新的发明。这种转变使建筑的发展成为可能,包括许多巨大的宫殿、寺庙和剧院,很多著名的古遗址因此而被世人所知。

以宏伟的埃及金字塔和美丽的希腊寺庙为例。农业的进步允许部分人成为科学家,研究天文学,由此开始了后来被用来探索世界的航海技术的发展。纵观历史,如果没有农业的发展,主要的人类发展都不会成为可能。农业的发展生产出可以养活艺术家、建筑工人、哲学家和科学家的多余食物。

后来,十八世纪末工业革命引起的城镇和城市的快速增长,使农业独立形成自己的区域。随着中耕机、收割机、脱粒机、联合收割机的发明,现代农业进一步发展。这些新技术使大型农业得以发展。现代科学也彻底改变了食品加工,如制冷的发明。今天,几乎每一种作物都是机械化收割。育种技术也使畜牧业生产更加高效。基因工程使种植农作物和饲养牲畜发生了变革。农业在使人们拥有自己喜欢的生活方式和自由方面扮演了极其重要的角色。

Lesson 3 Agriculture in China

Agriculture is a **vital industry** in China, employing over 500 million farmers. In 2019, China **ranked** first in worldwide farm **output**, primarily producing rice, wheat, potatoes, **sorghum**, **peanuts**, tea, **millet**, **barley**, **cotton**, **oilseed**, pork, and fish. Although **accounting** for only 10 percent of **arable** land worldwide, it produces food for 18 percent of the world's **population**.

Beginning of China's Agriculture

Beginning in about 7,500 BCE with **classical** millet agriculture, China's development of farming over the course of its history has played a key role in supporting the growth of what is now the largest population in the world. Jared Diamond estimated that the earliest **attested domestication** of rice took place in China by 7,500 BCE Excavations at Kuahuqiao, the earliest known Neolithic site in eastern China, have documented rice cultivation 7,700 years ago. Findings at the ruins of the Hemudu Culture in Yuyao and Banpo Village near Xi'an, which all date back 6,000 to 7,000 years, include rice, millet, and **spade-like** farm tools made of stone and bone.

Farming Method Improvements

Due to China's status as a developing country and its **severe** shortage of arable land, farming in China has always been very **labor-intensive**. However, throughout its history various methods have been developed or imported that enabled greater farming production and efficiency. They also **utilized** the **seed drill** to help improve on **row** farming.

During the Spring and Autumn Period (722-81 BCE), two **revolutionary** improvements in farming technology took place. One was the use of cast iron tools and beasts of burden to pull plows, and the other was the large-scale harnessing of rivers and development of water conservation projects. The engineer Sunshu Ao of the 6th century BCE and Ximen Bao of the 5th century BCE are two of the oldest **hydraulic** engineers from China, and their works were focused upon improving **irrigation** systems. These developments were widely spread during the ensuing Warring States Period (403-221 BCE), **culminating** in the enormous Dujiangyan Irrigation System engineered by Li Bing by 256 BCE for the State of Qin in ancient

Sichuan.

For agricultural purposes the Chinese had invented the hydraulic-powered **trip hammer** by the 1st century BCE. Although it found other purposes, its main function to **pound**, **decorticate**, and **polish** grain that otherwise would have been done manually. The Chinese also innovated the **square-pallet chain pump** by the 1st century AD, powered by a waterwheel or an oxen pulling on a system of mechanical wheels.

International Trade

China is the world's largest importer of soybeans and other food crops. China is expected to become the top importer of farm products within the next decade.

While most years China's agricultural production is sufficient to feed the country, in down years, China has to import grain. Due to the shortage of available farm land and an abundance of labor, it might make more sense to import land-extensive crops (such as wheat and rice) and to save China's scarce cropland for high-value export products, such as fruits, nuts, or vegetables. In order to maintain grain independence and ensure food security, however, the government of China has enforced policies that encourage grain production at the expense of more-profitable crops. Despite heavy restrictions on crop production, China's agricultural exports have greatly increased in recent years.

Vocabulary

vital [ˈvaɪtl] *adj.* 至关重要的
industry [ˈɪndəstri] *n.* 工业；产业
rank [ræŋk] *v.* 排列，属于某个等级
output [ˈaʊtpʊt] *n.* 产量
sorghum [ˈsɔːgəm] *n.* 高粱
peanut [ˈpiːnʌt] *n.* 花生
millet [ˈmɪlɪt] *n.* 小米，粟
barley [ˈbɑːli] *n.* 大麦
cotton [ˈkɒtn] *n.* 棉
oilseed [ˈɔɪlsiːd] *n.* 油料种子
account [əˈkaʊnt] *v.* （数量上，比例上）占
arable [ˈærəbl] *adj.* 适于耕种的
population [ˌpɒpjuˈleɪʃn] *n.* 人口
classical [ˈklæsɪkl] *adj.* 传统的

attest [əˈtest] *v.* 证实,证明
domestication [dəˌmestɪˈkeɪʃn] *n.* 驯养,驯化
spade-like [ˈspeɪd laɪk] *adj.* 像铲子的
severe [sɪˈvɪə(r)] *adj.* 严峻的;严厉的
labor-intensive [ˈlæbərɪntˈensɪv] *adj.* 劳动密集型的
utilize [ˈjuːtəlaɪz] *v.* 利用,使用
seed drill 条播机
row [rəʊ] *n.* 行,排
revolutionary [ˌrevəˈluːʃənəri] *adj.* 革命的
hydraulic [haɪˈdrɔːlɪk] *adj.* 水力的
irrigation [ˌɪrɪˈgeɪʃn] *n.* 灌溉
culminate [kʌlˈmɪneɪt] *v.* 达到顶点
trip hammer 杵锤
pound [paʊnd] *v.* 连续敲打
decorticate [diːˈkɔːtɪkeɪt] *vt.* 剥皮;去皮
polish [ˈpɒlɪʃ] *v.* 擦光
square-pillet *n.* 方坯
chain [tʃeɪn] *n.* 链子,链条
pump [pʌmp] *n.* 泵

Notes

1. Diamond, Jared 贾雷德·戴蒙德(Jared Diamond,1937 年 9 月 10 日—),美国演化生物学家、生理学家、生物地理学家以及非小说类作家。他最著名的作品《枪炮、病菌与钢铁》发表于 1997 年,获 1998 年美国普利策奖和英国科普图书奖。
2. Jared Diamond estimated that the earliest attested domestication of rice took place in China by 7,500 BCE Excavations at Kuahuqiao, the earliest known Neolithic site in eastern China, have documented rice cultivation 7,700 years ago.
   贾雷德·戴蒙德估计在中国,史料已证明的最早的农业种植水稻出现在发掘出的公元前 7 500年的跨湖桥遗址,位于中国东部,是现今已知最早的新石器时代遗址,在距今 7 700 年前就有文献记录了水稻种植。
3. One was the use of cast iron tools and beasts of burden to pull plows, and the other was the large-scale harnessing of rivers and development of water conservation projects.
   一次是使用铸铁工具和牲畜来拉犁,另一次是大规模地治理河流和发展水利工程。

Exercises

Ⅰ. Terms.

A. Put the following terms into Chinese.

1. Warring States Period
2. the square-pallet chain pump
3. the hydraulic-powered trip hammer
4. the Spring and Autumn Period
5. waterwheel
6. pipe systems
7. the chain pump
8. land-extensive crops
9. at the expense of
10. crop production

B. Put the following terms into English.

1. 高粱	6. 播种机
2. 花生	7. 杵锤
3. 棉花	8. 龙骨水车
4. 油料	9. 耕地
5. 遗址	10. 人口

Ⅱ. Fill in the blanks with the words given below. Change the forms where necessary.

climate	transport	fertile	concept	generation
conservation	critical	derive	livestock	variety

1. This is a ________________ moment.
2. Marketing people should pay more attention to this ________________.
3. Brazil has tried to balance development and ________________.
4. Marine ecosystems lost their regulatory abilities, and changes in ________ and weather followed.
5. She knew that she had to finally ______________ wisdom from this experience.
6. Is this the seventh __________ iPad?
7. But restrictions create ____________ ground for rumour-mongering.
8. This change in shape affects the cell's ability to ____________ oxygen.
9. __________ also plays a part in global warming.
10. There is a great ____________ of food in China.

Ⅲ. Complete the following passage with proper words.

While most years China's agricultural production is 1.________ to feed the country, in down years, China has to 2. ________ grain. Due to the shortage of available farm 3.________ and an abundance of labor, it might make 4.________ sense to import land-extensive crops (such as wheat and rice) and to 5.________ China's scarce cropland for high-value export products, such as, 6. ________, nuts, or vegetables. In order to maintain grain independence and ensure food 7. ________, however, the government of China has enforced policies that 8. ________ grain production at the 9. ________ of more-profitable crops. Despite heavy restrictions on crop production, China's agricultural exports have greatly 10. ________ in recent years.

Ⅳ. Translate the following sentences.

1. Although accounting for only 10 percent of arable land worldwide, it produces food for 20 percent of the world's population.

2. Due to China's status, as a developing country and its severe shortage of arable land, farming in China has always been very labor-intensive.

3. China is the world's largest importer of soybeans and other food crops.

4. Despite heavy restrictions on crop production, China's agricultural exports have greatly increased in recent years.

参考译文

中国的农业

在中国，农业是一个至关重要的产业，截至2019年，我国共有5亿多农民。中国农业产量排名全球第一，主要生产水稻、小麦、马铃薯、高粱、花生、茶叶、大麦、小米、棉花、油料、猪肉和鱼。虽然只占世界耕地面积的10%，却生产出了世界18%的人口所需的食品。

中国农业的开端

由大约公元前7 500年的古典粟农业开始，中国的农业在其历史发展进程中对目前人口居于世界首位的人口增长中起到了关键作用。贾雷德·戴蒙德估计在中国,史料已证明的最早的农业种植水稻出现在发掘出的公元前 7 500 年的跨湖桥,这是位于中国东部,现今已知最早的新石器时代遗址,在距今 7 700 年前就有文献记录了水稻种植。在西安附近发现的距今 6 000 至 7 000 年前的余姚和半坡村的河姆渡文化遗址中出现了大米、小米和用石头和骨头制成的铲等农具。

耕作方法的改进

由于中国是一个发展中国家,耕地严重短缺,中国农业一直都是劳动密集型的。然而,在其整个历史上已开发或引进了各种不同的方法来增加农业生产,提高生产效率。他们还使用播种机来帮助提高种植。

在春秋时期(公元前 722—81 年),发生了两次农业技术的革新。一次是使用铸铁和牲畜来拉犁,另一次是大规模地治理河流和发展水利工程。公元前六世纪的工程师孙叔敖,公元前五世纪的西门豹是中国两个最古老的水利工程师,他们的工作主要集中在改善灌溉系统。在随后的战国时期(公元前 403—221 年),这些成果被广泛传播,最终在公元前 256 年由李兵为秦国在古时的四川设计建造巨大的都江堰系统灌溉工程时达到顶峰。

从农业的目的出发,中国人在公元前一世纪发明了水力铁锤。虽然也有其他的目的,其主要的功能为碾碎、剥皮、磨光粮食,否则这些就要手工完成。公元一世纪中国人也革新了龙骨水车,该系统由水车或用牛牵引拉机械齿轮提供动力。

国际贸易

中国是世界上最大的大豆和其他粮食作物的进口国。未来十年内,中国有望成为最大的农产品进口国。

虽然大多数年份里,中国的农业生产足以养活整个国家,在产量低的年份,中国必须进口粮食。由于现有的耕地和劳动力的短缺,进口集约作物(如小麦和大米),保护稀缺的耕地。种植经济价值高的作物,如水果、坚果、蔬菜,对中国可能更有意义。然而,为了保持粮食的独立性,确保粮食安全,中国政府已实施政策,鼓励粮食生产而不是经济作物生产。尽管对作物生产有严格限制,但是近几年来,中国农产品出口已经大大增加。

Unit 2
Plants

The trees have green and the earth has a pulse.
——树木拥有绿色,地球才有脉搏。

Situational Dialogue:

Hydroponics on Flowers

A: Excuse me, could you please tell me what these flowers are called?

B: Yes, they're called water-cultured flowers.

A: Is it that these flowers are frown in water rather than in soil?

B: Quite right. The soil-less and sexless reptoduction technology is used. A leaf blade can be used as the cultured tissue.

A: What are the advantages?

B: Well, no special equipment and fertilizer are required and this makes the culture cost cheaper, and the culture coefficient higher.

A: How long will it take for the root to spring up?

B: Not very long, about 3-7days.

A: How much investment is needed?

B: An investment from US \$1,000 to \$2,000 is a minimum.

A: What is the profit rate?

B: It can reach as much as 200%.

Lesson 1 Starting to Grow

Where Do Plants Come From?

Plants come from seeds. Each seed contains a **tiny** plant waiting for the right conditions to germinate, or start to grow.

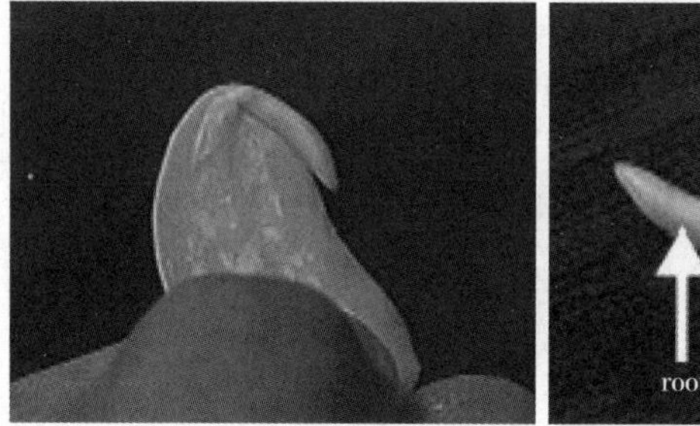

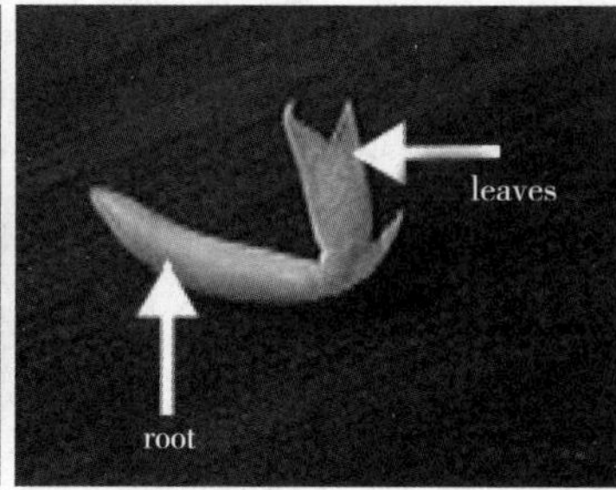

What Do Seeds Need to Start to Grow?

Seeds wait to germinate (to germinate means to start to grow) until three needs are met: water, correct temperature (warmth), and a good **location** (such as in soil). During its early stages of growth, the seedling relies upon the food **supplies** stored with it in the seed until it is large enough for its own leaves to begin making food through photosynthesis. The seedling's roots push down into the soil to **anchor** the new plant and to **absorb** water and minerals from the soil. And its **stem** with new leaves pushes up toward the light.

The germination **stage** ends when a **shoot emerges** from the soil. But the plant is not done growing. It's just started. Plants need water, warmth, **nutrients** from the soil, and light to **continue** to grow.

What Do Different Plant Parts Do?

Plant parts do different things for the plant.

Roots

Roots act like **straws** absorbing water and **minerals** from the soil. Tiny **root hairs stick** out of the root, helping in the absorption. Roots help to anchor the plant in the soil so it does not fall over. Roots also store extra food for future use.

Stems

Stems do many things. They support the plant. They act like the plant's **plumbing system**, **conducting** water and nutrients from the roots and food in the form of **glucose** from the leaves to other plant parts. Stems can be **herbaceous** like the **bendable** stem of a **daisy** or **woody** like the **trunk** of an **oak** tree.

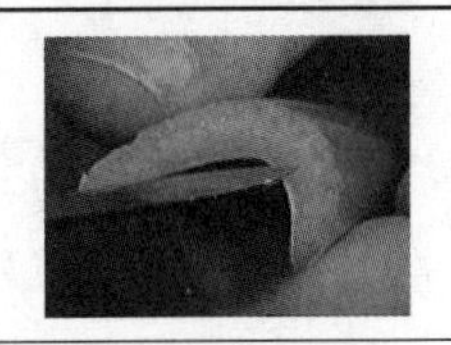	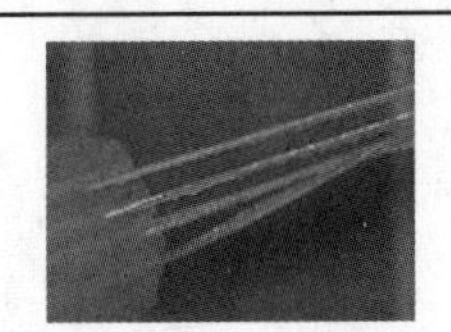	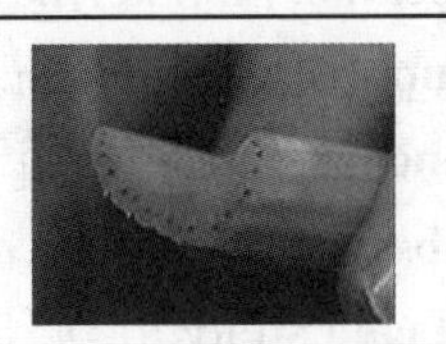
A **celery stalk**, the part of celery that we eat, is a special part of the leaf **structure** called a **petiole**. A petiole is a small stalk **attaching** the leaf **blade** of a plant to the stem.	In celery, the petiole serves many of the same functions as a stem. It's easy to see the "pipes" that conduct water and nutrients in a stalk of celery.	Here you can easily see the "pipes".

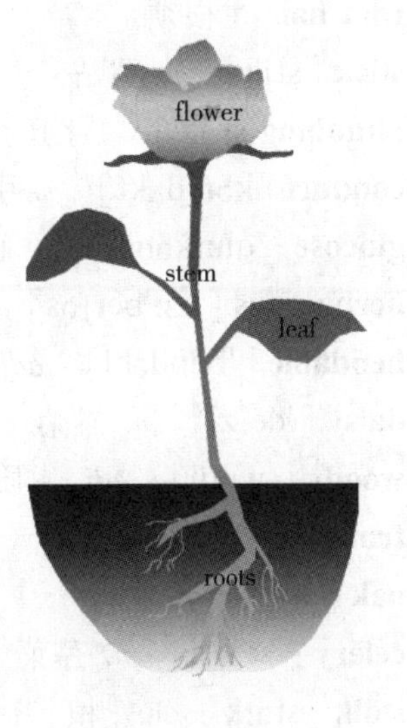

Leaves

Most plants' food is made in their leaves. Leaves are designed to **capture** sunlight which the plant uses to make food through a process called photosynthesis.

Flowers

Flowers are the **reproductive** part of most plants. Flowers contain **pollen** and tiny eggs called **ovules**. After **pollination** of the flower and **fertilization** of the ovule, the ovule develops into a fruit.

Fruit

Fruit provides a covering for seeds. Fruit can be fleshy like an apple or hard like a nut.

Seeds

Seeds contain new plants. Seeds form in fruit.

What's the Difference Between a Fruit and a Vegetable?

A fruit is what a flower becomes after it is pollinated. The seeds for the plant are inside the fruit.

Vegetables are other plant parts. Carrots are roots. **Asparagus** stalks are stems. Lettuce is leaves.

Foods we often call vegetables when cooking are really fruits because they contain seeds inside.

Vocabulary

tiny [ˈtaɪni] *adj.* 极小的
location [ləʊˈkeɪʃn] *n.* 位置,场所
supply [səˈplaɪ] *n.* 供给物
anchor [ˈæŋkə(r)] *v.* 抛锚
absorb [əbˈsɔːb] *v.* 吸收
stage [steɪdʒ] *n.* 阶段
shoot [ʃuːt] *n.* 幼苗,嫩芽
emerge [iˈmɜːdʒ] *v.* 出现,浮现；暴露
nutrient [ˈnjuːtriənt] *n.* 营养物,营养品
continue [kənˈtɪnjuː] *v.* 持续
straw [strɔː] *n.* 稻草；麦秆
mineral [ˈmɪnərəl] *n.* 矿物质
root hair 根须
stick [stɪk] *v.* 粘贴
plumbing system 管道系统
conduct [kənˈdʌkt] *v.* 引导
glucose [ˈgluːkəʊs] *n.* 葡萄糖
herbaceous [hɜːˈbeɪʃəs] *adj.* 草本的
bendable [ˈbendəbl] *adj.* 可弯曲的
daisy [ˈdeɪzi] *n.* 雏菊
woody [ˈwʊdi] *adj.* 木质的
trunk [trʌŋk] *n.* 树干
oak [əʊk] *n.* 栎树
celery [ˈseləri] *n.* 芹菜
stalk [stɔːk] *n.* [植]茎,秆
structure [ˈstrʌktʃə(r)] *n.* 结构
petiole [ˈpetɪəʊl] *n.* 叶柄,柄部
attach [əˈtætʃ] *v.* 贴上,系
blade [bleɪd] *n.* (壳、草等的)叶片
capture [ˈkæptʃə(r)] *v.* 俘获；夺取
reproductive [ˌriːprəˈdʌktɪv] *adj.* 生殖的；再生产的
pollen [ˈpɒlən] *n.* 花粉
ovule [ˈɒvjuːl] *n.* 胚珠,卵子
pollination [ˌpɒləˈneɪʃn] *n.* [植]授粉(作用)

fertilization [ˌfɜːtəlaɪˈzeɪʃn] *n.* 施肥
asparagus [əˈspærəgəs] *n.* 芦笋
lettuce leave 生菜叶

1. herbaceous：草本的
   Plants with stems that are usually soft and bendable. Herbaceous stems die back to the ground every year.
   草本植物的茎柔软、易弯曲。草本植物为一年生。
2. woody：木质的
   Plants with stems, such as tree trunks, that are hard and do not bend easily, usually don't die and live back to the ground each year.
   木本植物如树木，其茎坚硬而不易弯曲，通常为多年生植物。
3. photosynthesis：光合作用
   A process by which a plant produces its food using energy from sunlight, carbon dioxide from the air, and water and nutrients from the soil.
   光合作用是植物运用阳光中的能量、空气中的二氧化碳以及土壤中的水和矿物质，制造营养物质的过程。
4. pollination：授粉
   The movement of pollen from one plant to another. Pollination is necessary for seeds to form in flowering plants.
   花粉从一株植物到另一株植物的移动。开花植物的授粉对种子的形成是必须的。
5. Seeds wait to germinate (to germinate means to start to grow) until three needs are met：water, correct temperature (warmth), and a good location (such as in soil).
   种子发芽(发芽就是开始生长)有三个必备要素：水、合适的温度和适宜的地方(例如土壤)。
   are met 是被动语态；meet needs：满足需求。
6. During its early stages of growth, the seedling relies upon the food supplies stored with it in the seed until it is large enough for its own leaves to begin making food through photosynthesis.
   生长初期，幼苗完全依靠存储在种子里的营养生长，等到它长出叶子，通过光合作用，才能够给自身提供营养物质。
   the seedling relies upon the food supplies stored with it in the seed 是主句，the seedling 是主句中的主语；relies upon 是谓语，"依赖，依靠"；the food supplies 是宾语，stored with it in the seed 是宾语补足语；until 引导时间状语从句。
7. They act like the plant's plumbing system, conducting water and nutrients from the roots and food in the form of glucose from the leaves to other plant parts.

它们好似植物的管道系统,从植物的根部运输水和营养物质,并从叶子中以葡萄糖的形式为植物其他部分提供食物。

in the form of 以……的形式

8. Leaves are designed to capture sunlight which the plant uses to make food through a process called photosynthesis.

叶子用来吸收阳光,有了阳光,植物就可以通过光合作用的过程生成食物。

are designed 是被动语态,意为"被设计",表示客观性。

9. After pollination of the flower and fertilization of the ovule, the ovule develops into a fruit.

花授粉和胚珠受精后,胚珠发育为果实。

develop into 发展成,发育为……

10. Foods we often call vegetables when cooking are really fruits because they contain seeds inside.

烹饪时我们常常称之为蔬菜的食物事实上是果实,因为它们包含了植物的种子。

Exercises

Ⅰ. Supermarket botany.

People eat many different parts of plants. We all know that an apple is a fruit—it contains the apple tree's seeds. But can you guess which plant part each food below is?

A. Root B. Stem C. Leaf D. Seed E. Flower F. Fruit

Asparagus	Broccoli	Carrots
( )	( )	( )
Coffee	**Peach**	**Lettuce**
( )	( )	( )

续表

Celery	Cucumber	Radish
(　　)	(　　)	(　　)
Green Pepper	Sweet Potato	Spinach
(　　)	(　　)	(　　)
Orange	Peanuts	Tomatoes
(　　)	(　　)	(　　)

Ⅱ. Answer the following questions. If necessary, try to find some resources in your library or on the Internet.

1. Where do plants come from?

2. What do seeds need to start to grow?

3. What do different plant parts do?

4. What's the difference between a fruit and a vegetable?

Ⅲ. Translate the following short passage into Chinese.

Leaf Printing

Paint one side of the leaf. The imprint will show up better if you paint the bottom side of the leaf where the veins stick out more.

Put newspaper under the fabric or between the fabric layers if you're painting on a T-shirt. Lay the painted leaf on the fabric (cotton works well) or paper and apply equal pressure to all parts of the leaf. This method is call pressure printing. With a little practice you'll discover how hard to press the leaf and how much paint to apply.

A rolling pin sometimes makes this process easier. A closet rod, cut into 12 inch lengths mankes an inexpensive set of rolling pins. You can also use a printing technique. Lay your printed leaf down, paint side up, and lay the material to be printed on top of the leaf. In one motion firmly press down on the material and leaf.

Ⅳ. Wrting.

With increasing awareness of environment, we realized how important plants are to our human beings. Can you introduce some functions of plants?

参考译文

生　长

植物从何而来?

植物是由种子生长来的。每粒种子都包含一个细小的物质,在适当的条件下它们发芽或者开始生长。

种子生长需要什么?

种子发芽(发芽就是开始生长)有三个必备要素:水、合适的温度和适宜的地方(例如土壤)。生长初期,幼苗完全依靠存储在种子里的营养生长,等到它长出叶子,通过光合作用,才能够给自身提供营养物质。为了固定新生植物,从土壤里吸收水和营养,幼苗的根向下延伸到土壤深处。长出新叶的茎则迎着阳光向上生长。

幼苗从土壤中长出的时候萌芽阶段结束。但这不是植物生长的结束。这才是生长的开始。植物需要水,温度,土壤中的营养物质和光继续生长。

植物不同部分因何不同?

植物的不同组成部分发挥不同作用。

根

如同稻草一样,根从土壤里吸收水和矿物质。细小的根须扎根在泥土里,有助于吸收营养。根帮助植物在土壤中直立稳固而不至于倒下。同时,根也为未来存储养分。

茎

茎也有许多功能。它们支撑植物。它们好似植物的管道系统,从植物的根部运输水和营养物质,并从叶子中以葡萄糖的形式为植物其他部分提供食物。茎既可以是雏菊一样可以弯曲的草本茎,也可以是橡树树干一样的木质茎。

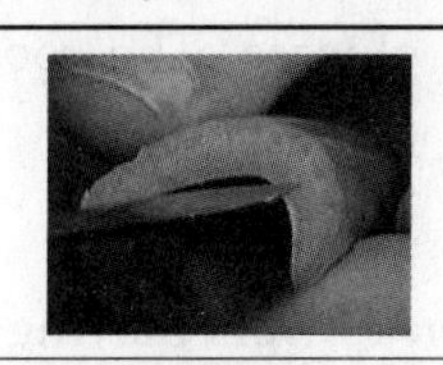	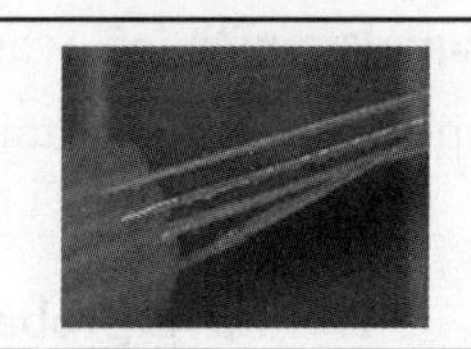	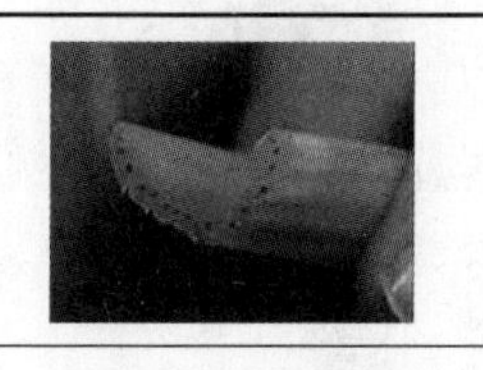
芹菜茎,就是我们吃的芹菜的那一部分,是被称为叶柄的特殊叶片结构。叶柄是连接植物叶子与茎的那一小部分。	芹菜的叶柄有许多茎的相同功能。在芹菜的茎干里很容易看到运输水分和营养的“管道”。	这张图片里,你可以很容易看到这些“管道”。

叶子

大多数植物的食物都是在叶子里产生的。叶子用来吸收阳光,有了阳光,植物就可以通过光合作用的过程生成食物。

花

花是大多数植物的生殖器官。花含有花粉和叫胚珠的微小的受精卵。花授粉和胚珠受精后,胚珠发育为果实。

果

果实包裹着种子。果实可以是像苹果那样的果肉,或者是像核桃那样的硬壳。

种子

种子里包含着新生植物。种子在果实内形成。

果实与蔬菜的区别

果实是花授粉之后长成的。植物的种子长在果实里。

蔬菜是植物的其他部分。胡萝卜是根。芦笋是茎。生菜是叶。

烹饪时我们常常称之为蔬菜的食物事实上是果实,因为它们包含了植物的种子。

Lesson 2 The Parts of a Vascular Plant

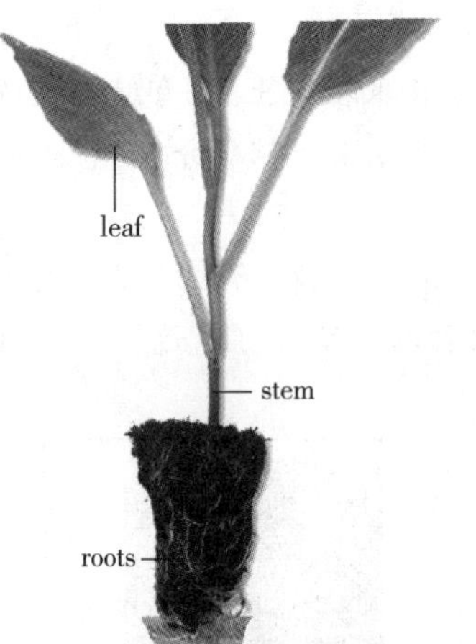

What **anchors** a plant and absorbs **minerals** and water from soil? That is the job of the plant's roots. What do leaves do? The plant's leaves use water to make food in the form of **glucose**. Why is the **stem** important? A plant's stem supports the plant and transports water, minerals and glucose.

A vascular plant has tiny tubes that transport **liquids**, such as water and glucose. The roots, leaves and stems all contain these tiny tubes.

Roots

You have probably seen flowers moving when the wind blows. Why don't they blow away? The flowers are held in place by their **underground** roots. Roots help the plant stay secure in the ground. They start from the **base** of the plant and **spread** in the soil. There are many types of roots.

Some plants have only one main root, with many tiny roots growing from it. This type of root is called **taproot**. **Dandelions** have taproots. Other plants have many smaller roots that spread out like tiny arms.

Most roots are in the soil, which contains water and minerals. Water and minerals enter the root through a thin outer layer of **cells** called the **epidermis**. The epidermis has **root hairs** that help the root absorb water from the soil. If the epidermis did not have root hairs, it would absorb less water. Nutrients move from the epidermis to the **xylem**. The xylem, a type of vascular **tissue**, moves water and minerals from the roots to other plant parts. Some plants, such as certain **orchids**, have **aerial** roots, which **extract moisture** from the air.

Plants make their own food. The leaves of a plant make a type of **sugar** called glucose. Another type of vascular tissue is **phloem**, which moves glucose to the different parts of the plant. Roots store some of this glucose in the form of **starch**. Beets are roots that store food.

Stems

Stems come in different forms. They can be tall or short, **rough** or **smooth**, **curved** or straight. Stems support plants. Like roots, they have xylem and phloem. The xylem and phloem move water and glucose between the roots and the leaves of the plant.

Some stems are green and **wary** to bend. Plants with this type of stem are called **herbaceous** plants. The leaves and stems of these plants can die in cold weather, but their roots keep on living underground. Each year, these plants grow new stems. **Strawberries**, grasses, and weeds are herbaceous plants.

Other stems are strong and thick. Plants with this type of stem are called **woody** plants. These plants grow to be large and live for a long time. A woody plant may lose its leaves for part of the year, but its stem remains alive. Examples of woody plants are trees, **shrubs** and **vines**.

There are also stems that grow under the ground. Potatoes are an example of underground stems. Food stored in **tubers** helps the plant to survive. For example, if it does not rain enough or it is too cold, a plant may not make enough sugar. It survives because it has a store of food in its tubers.

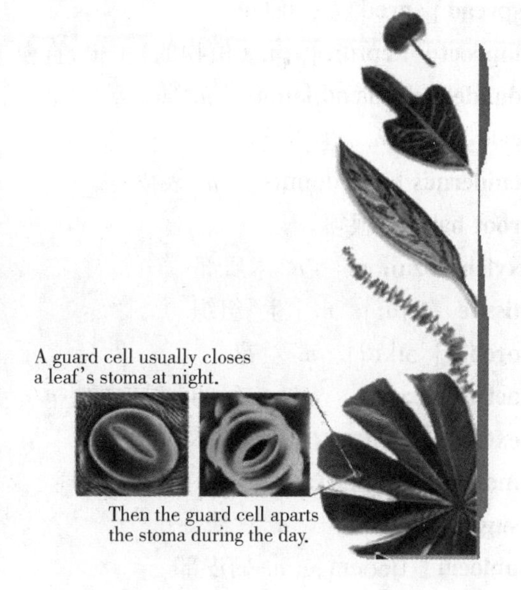
A guard cell usually closes a leaf's stoma at night.

Then the guard cell aparts the stoma during the day.

Leaves

Leaves come in all shapes and sizes. But the function they all have in common is

that they make food for plants. As leaves make a food called **glucose**, water and gases travel in and out of the plant through tiny holes in the leaf's **epidermis**. Each hole is called a **stoma** (plural: stomata). The **cell** that opens and closes a leaf's **stoma** is a guard **cell**.

Sunlight can cause guard cells to take in water. The water that is taken in through the guard cells puts pressure on their walls. This increased pressure causes the shape of the guard cells to become **curved**. Once the guard cells curve, the stomata open.

Vocabulary

vascular [ˈvæskjələ(r)] *adj.* 脉管的
anchor [ˈæŋkə(r)] *v.* 抛锚
mineral [ˈmɪnərəl] *n.* 矿物质
glucose [ˈgluːkəʊs] *n.* [化]葡萄糖
stem [stem] *n.* (花草的)茎
liquid [ˈlɪkwɪd] *n.* 液体
underground [ˈʌndəgraʊnd] *adj.* 地下的
base [beɪs] *n.* 基础
spread [spred] *v.* 伸开
taproot [ˈtæpruːt] *n.* (植物的)主根,直根
dandelion [ˈdændɪlaɪən] *n.* 蒲公英
cell [sel] *n.* [生]细胞
epidermis [ˌepɪˈdɜːmɪs] *n.* 表皮
root hairs 根须
xylem [ˈzaɪləm] *n.* 木质部
tissue [ˈtɪʃuː] *n.* [生]组织
orchid [ˈɔːkɪd] *n.* 兰花
aerial [ˈeəriəl] *adj.* 空气的；航空的 *n.* [电讯] 天线
extract [ˈekstrækt] *v.* 提取
moisture [ˈmɔɪstʃə(r)] *n.* 水分；湿气
sugar [ˈʃʊgə(r)] *n.* 食糖
phloem [ˈfləʊem] *n.* 韧皮部
starch [stɑːtʃ] *n.* 淀粉
rough [rʌf] *adj.* 粗糙的
smooth [smuːð] *adj.* 光滑的
curved [kɜːvd] *adj.* 弧形的,弯曲的

wary [ˈweəri] *adj.* 谨慎的，小心翼翼的
herbaceous [hɜːˈbeɪʃəs] *adj.* 草本的
strawberry [ˈstrɔːbəri] *n.* 草莓
woody [ˈwʊdi] *adj.* 木质的；木本的
shrub [ʃrʌb] *n.* 灌木；灌木丛
vine [vaɪn] *n.* 藤；藤本植物
tuber [ˈtjuːbə(r)] *n.* （植物的）块茎；结节

Notes

1. vascular plant：维管植物。现存的维管植物有25万~30万种，包括极少部分苔藓植物、蕨类植物（松叶兰类、石松类、木贼类、真蕨类）、裸子植物和被子植物。维管系统（木质部和韧皮部）的产生是植物从水生到陆生长期适应环境的结果。维管系统的有效输导，使维管植物成为最繁茂的陆生植物。较原始的维管植物木质部中多只具管胞，故也可称这些植物为管胞植物。
2. What anchors a plant and absorbs minerals and water from soil?
   植物靠什么固定，又是什么使植物从土壤中吸收矿物质和水？
   absorb：*v.* 吸收（液体、气体等）。如：
   Dry earth absorbs water quickly.
   干土吸水很快。
   Over the centuries, they gradually absorbed Islamic ideas about design and architecture.
   几个世纪以来，他们渐渐汲取了伊斯兰教在设计和建筑方面的理念。
   The task absorbed his time.
   那任务占用了他的时间。
3. in the form of 以……的形式，如：
   The company also likes to spread a little holiday cheer in the form of annual bonuses.
   公司还以年度资金的形式让员工的节日过得更加开心。
4. A plant's stem supports the plant and transports water, minerals, and glucose.
   植物的茎支撑它，并为它运输水、矿物质和葡萄糖。
   support：*v.* 支持；帮助；支撑，维持。如：
   Is there revenue to support the business?
   有没有收入支撑这项业务？
5. Another type of vascular tissue is phloem, which moves glucose to the different parts of the plant.
   另一种类型的维管组织是韧皮部，可把葡萄糖运送到植物的不同部位。

Exercises

Ⅰ. Read the text and decide whether each of the following statements is True（T）or False（F）.

1. A plant's leaf supports the plant and transports water, minerals and glucose.

2. All roots are in the soil, which contains water and minerals.

3. Some plants have only one main root, with many tiny roots growing from it.

4. Roots store some of this glucose in the form of starch.

5. Trees, shrubs and vines are herbaceous plants.

6. A woody plant may lose its leaves for part of the year, but its stem remains alive. For example, strawberries, grasses and weeds.

Ⅱ. Read the text and answer the following questions.

1. What anchors a plant and absorbs minerals and water from soil?

2. What do leaves do? What is the function of leaves?

3. Why is the stem important?

4. You have probably seen flowers moving when the wind blows. Why don't they blow away?

5. How many types of roots are there? What are they?

Ⅲ. Translate the following paragraph.

Chinese agriculture has to undergo a low-carbon development with characteristics of being resources-saving, production-clean, environment-friendly and quality-efficiency-oriented. Therefore, in terms of technology, great efforts must be made to develop a series of key technologies and promote their practical application. For example, resources-saving technology including energy-saving, soil-saving, water-saving, fertilizer-saving, insecticides-saving, seeds-saving, materials-saving and labor-saving techniques, and so on, technology of reducing the use of agricultural chemicals and developing their substitutes, technology of cultivating new plant species with a high light absorption and carbon sequestration, technology of sequestratrating soil carbon, technology of developing clean energies, clean environment-friendly production technology, technology of nuisance-free proposal and reclamation of wastes, and so on.

参考译文

维管植物的组成

植物靠什么固定,又是什么使植物从土壤中吸收矿物质和水?这些都是靠植物根的工作。树叶利用水形成葡萄糖形式的食物。为什么茎很重要?植物的茎支撑它,并为它运输水、矿物质和葡萄糖。

维管植物有微小的输送如水和葡萄糖等液体的管。根、叶、茎都有这些细小的管子。

根

你可能会看到当风吹起的时候,花在移动。它们为什么不会被吹走?花是由长在地下的根支撑的。根帮助植物扎根泥土。它们从植物的根开始生长,慢慢地根系在土壤中蔓延散开。根有许多类型。

有些植物只有一个主根,还有许多从主根上长出的微小的根。这种类型的根被称为主根。蒲公英有主根。其他植物有许多像小胳膊一样蔓延开来的较小的根。

大多数根长在土壤里,土壤中含有水和矿物质。水和矿物质通过一个薄的被称为表皮的外层进入根。表皮细胞有根须可以帮助根从土壤中吸收水分。如果表皮细胞没有根毛,它吸收的水就更少。营养物质从表皮到木质部。木质部是一个维管组织,使水和矿物质从根部到植物的其他部分。有些植物如某些兰花,有气根,可以从空气中提取水分。

植物可以自己制造食物。植物的叶子制造出一种称为葡萄糖的糖。另一种类型的维管组织是韧皮部，可把葡萄糖运送到植物的不同部位。植物的根以淀粉的形式存储这种葡萄糖。甜菜的根就是这种存储食物的根。

茎

茎有不同的形式。它们可高可矮，可粗糙可光滑，可弯曲可笔直。茎支撑着植物。和根一样，它们也有木质部和韧皮部。木质部和韧皮部在植物的根和叶之间运输水和葡萄糖。

一些茎是绿色的、易弯曲，有这种茎的植物被称为草本植物。这些植物的茎和叶在寒冷的天气会死亡，但它们的根继续生活在地下。每年这些植物生长出新的茎。草莓、草、杂草都是草本植物。

其他的茎很强壮。有这种茎的植物被称为木本植物。这些植物长得很大，可以长期生存。木本植物可能在一年的某一阶段落叶，但其树干仍然活着。木本植物的例子有树木、灌木、藤本。

还有一些茎生长在地下。土豆是一个茎长在地下的例子。存储在块茎中的食物帮助植物生存。例如，如果没有足够的雨水或天气太冷，植物可能制造不出足够的糖。这种植物仍然能够生存，因为它的块茎里储藏了食物。

叶

叶形状、大小各异。但它们都有共同的功能就是为植物提供食品。树叶生成称为糖、水和气体的食物，这些食物通过植物表皮的小孔在植物各部分运输。每个孔称为气孔（复数形式 stomata）。打开和关闭叶片气孔的是保卫细胞。

阳光可能会引起保卫细胞吸收水分。从保卫细胞吸收的水在细胞壁上增压。增加的压力导致保卫细胞形状变弯。一旦保卫细胞弯曲，气孔就会打开。

Lesson 3 What Do Plants Need?

To grow satisfactorily, a plant needs warmth, light, water, carbon **dioxide** and about a dozen other chemical **elements** which it can obtain from the soil.

Warmth

Most crop plants in this country start growing when the average daily temperature is above 6 ℃ (42 ℉). Growth is best between 16 ℃ (60 ℉) and 27 ℃ (80 ℉). These temperatures **apply** to **thermometer readings** taken in the shade about 4ft above ground. Crops grow in hotter countries usually have higher temperature **requirements**.

Cold **frosty** conditions may seriously damage plant growth. Crop plants **differ** in their ability to withstand very cold conditions. For example, winter **rye** and wheat can stand colder conditions than winter oats. Potato plants and stored tubers are easily damaged by frost. Sugar-beet may **bolt** (go to seed) if there are frosts after **germination**; frost in December and January may damage crops left in the ground.

Light

Without light, plants cannot produce **carbohydrates** and will soon die. The **amount** of **photosynthesis** which takes place daily in a plant is partly due to the length of daylight and partly to the **intensity** of the sunlight. Bright sunlight is of most importance where there is dense plant growth.

The lengths of daylight and darkness periods vary according to the distance from the **equator** and also from season to season. This can **affect** the flowering and seeding of crop plants and is one of the limiting factors in introducing new crops into a country. Grasses are now being tested in this country which will remain leafy and not produce flowering **shoots** under the daylight conditions here.

Water

Water is an essential part of all plant cells and it is also required in **extravagant** amounts for the process of **transpiration**. Water carries nutrients from the soil into and through the plant and also carries the products of photosynthesis from the leaves to wherever they are needed. Plants take up about 200 tons of water for every ton of dry matter produced.

Carbon Dioxide (CO_2)

Plants need carbon dioxide for photosynthesis. This is taken into the leaves through the **stomata** and so the amount which can go in is affected by the rate of transpiration. Another limiting factor is the small amount (0.03%) of carbon dioxide in the **atmosphere**. The percentage can increase just above the surface of soils rich in organic matter where soil bacteria are active and **releasing** carbon dioxide. This is possibly one of the reasons why crops grow better on such soils.

Chemical Elements

In order that a plant may build up its cell structure and **function** as a food factory, many simple chemical **substances** are needed. These are absorbed into the roots from the soil **solution** and the **clay particles**. Those required in fairly large amounts—a few kilogram to one or more hundred kilogram per **hectare** (a few pounds to one or more **hundredweight** per **acre**) are called the major **nutrients**; those required in small amounts—a few grams to several kilogram per hectare (part of an **ounce** to several pounds per acre) are the minor nutrients or **trace** elements.

Vocabulary

dioxide [daɪ'ɒksaɪd] *n.* 二氧化物
element ['elɪmənt] *n.* [化]元素
apply [ə'plaɪ] *v.* 应用,运用
thermometer [θə'mɒmɪtə(r)] *n.* 温度计；体温表
reading ['riːdɪŋ] *n.* 读数
requirement [rɪ'kwaɪəmənt] *n.* 要求；必需品
frosty ['frɒstɪ] *adj.* 严寒的；霜冻的
differ ['dɪfə(r)] *v.* 使……不同
rye [raɪ] *n.* 黑麦,裸麦
bolt [bəʊlt] *v.* 迅速成长并结实
germination ['dʒɜːmɪ'neɪʃn] *n.* 萌芽
carbonhydrate ['kɑːbəʊ'haɪdreɪt] *n.* 碳水化合物
amount [ə'maʊnt] *n.* 总量 *v.* 合计,总共
photosynthesis ['fəʊtəʊ'sɪnθəsɪs] *n.* 光合作用
intensity [ɪn'tensətɪ] *n.* 强度；剧烈
equator [ɪ'kweɪtə(r)] *n.* 赤道
affect [ə'fekt] *v.* 影响
shoot [ʃuːt] *n.* 嫩枝,苗
extravagant [ɪk'strævəgənt] *adj.* 过高的,过分的
transpiration ['trænspɪ'reɪʃn] *n.* 蒸腾作用
stoma ['stəʊmə] *n.*（复数 stomata）气孔
atmosphere ['ætməsfɪə(r)] *n.* 大气,空气
release [rɪ'liːs] *v.* 释放
function ['fʌŋkʃn] *v.* 起作用
substance ['sʌbstəns] *n.* 物质
solution [sə'luːʃn] *n.* 溶液
clay [kleɪ] *n.* 黏土；泥土
particle ['pɑːtɪkl] *n.* 粒子,微粒
hectare ['hekteə(r)] *n.* 公顷(缩略形式为 ha.)
hundredweight ['hʌndrədweɪt] *n.* 英担
acre ['eɪkə(r)] *n.* 英亩
nutrient ['njuːtriənt] *n.* 营养品,养料
ounce [aʊns] *n.* 盎司
trace [treɪs] *n.* 丝毫,微量

Notes

1. 6 ℃,读作 six degrees Centigrade,摄氏六度; 42 ℉读作 forty-two degrees Fahrenheit,华氏四十二度; 4ft,读作 four feet。
2. withstand [wɪð'stænd] *v.* 经受,承受,禁得起; 反抗。如:
   withstand the storm 顶得住暴风雨
   withstand severe tests 经得起严峻的考验
   withstand earthquake 耐震
   withstand wear 耐穿
3. go to seed=run to seed 花谢结籽
4. due to: 由于,应归于。它引导的介词短语在句中一般作表语或定语。如:
   The amount of photosynthesis which takes place daily in a plant is partly due to the intensity of the sunlight.
   植物每天发生的光合作用的量部分取决于日光的强度。
   These are errors due to carelessness.
   这些是由于粗心而引起的错误。
   All our achievements are due to the correct leadership of our Party.
   我们的一切成就都应归功于党的正确领导。
5. Grasses are now being tested in England which will remain leafty and not produce flowering shoots under the daylight conditions here.
   现在,在英国正在对一些禾本科植物进行试验,它们在这里的日光条件下总是叶子茂盛但却不长花枝。
   主语中的谓语 are being tested 为正在进行时的被动语态。which 引导的定语从句隔着主语谓语说明前面的主句主语 grassess。这种定语句称为"分裂式定语从句",常见于主语有被动语态谓语的情况中。如:
   A few stars are known, which are hardly bigger than the earth.
   已知的星球,大小和地球相仿的寥寥无几。
6. take up: 吸收(液体); 溶解(固体); 占(时间,空间); 接受; 接纳(乘客); 拿起; 继续; 学习。例句如下:
   Plants take up about 200 tons of water for every ton of dry matter prodeuced.
   植物每产生 1 吨干物质就需要吸收 200 吨水。
   How much water is needed to take up a pound of salt?
   溶解 1 磅盐需要多少水?
   My time is fully taken up with writing.
   写作占用了我的全部时间。
   The building takes up 400 square meters.

这栋楼占地400平方米。

7. rich in orgain matter 为“形容词+介词短语”构成形容词短语，作定语时，其位置一般置于所修饰的名词之后。又如：

   Water is a substance suitable for preparation of hydrogen and oxygen.

   水是适于制取氢和氧的物质。

8. in order that 为一连接词词组，意为“以便；为了”，用来引导目的状语从句。如：

   In order that a plant may build up its cell structure and function as a food factory many simple chemical substances are needed.

   植物为了建立其细胞结构和食品工厂的作用，就需要很多简单的化学物质。

9. build up：树立，逐步建立；积累（资金）；增进（健康）；集结；增加。如：

   A plant can build up its cell structure.

   植物能建立其细胞结构。

10. function as：起……的作用。如：

    A plant can function as a food factory.

    植物能起食品工厂的作用。

Exercises

Ⅰ. Match the items listed in the following columns.

1. 有机物质	A. carbon dioxide
2. 蒸腾作用	B. daily temperature
3. 土壤中的营养成分	C. thermometer
4. 发芽	D. photosynthesis
5. 二氧化碳	E. germination
6. 日平均温度	F. transpiration
7. 温度计	G. nutrients from the soil
8. 光合作用	H. organic matter
9. 微量元素	I. go to seed
10. 花谢结籽	J. the minor nutrients

Ⅱ. Read the text and answer following questions.

1. What does a plant need?

2. What is the best temperature for growth?

3. What is the function of light?

4. What is the function of water?

5. What is the function of carbon dioxide?

6. What is the function of chemical elements?

Ⅲ. Translate the following paragraph.

With the development of both economy and civilization in human society, people have greatly improved their knowledge and understanding about forests. They therefore have also dramatically changed their social demands for forestry. As a result, more world attention has been widely paid to the function that forests play to maintain and improve environment. In 1992, UN Conference on Environment and Development bestowed priority on forestry and it became a political promise of the highest rank. In addition, it was particularly emphasized in the meeting that nothing has been more important than forestry among the problems that the world summit conferences will deal with. It is a distinct milestone in world civilization history to place forestry issues at such a high position.

参考译文

植物需要什么？

为满足生长，植物需要温度、光照、水分、二氧化碳和十几种其他可以从土壤中获得的化学元素。

温度

在我国，当日平均温度高于6 ℃（42 ℉）时大多数农作物开始生长。温度在16 ℃（60 ℉）和27 ℃（80 ℉）之间时，作物长势最好。这些温度是将温度计置于地面以上4英尺的阴凉处测得的。在气温更高的国家，作物生长通常有更高的温度要求。

严寒的条件可能会严重损害植物生长。作物抵御严寒的能力不同。例如,冬黑麦和小麦能抵御比冬季燕麦更能抵御寒冷的条件。马铃薯植株和储存的块茎容易受严寒霜冻损坏。甜菜在发芽以后有霜冻可能抽薹(花谢结籽);十二月和一月的霜冻可能损害留在地面上的作物。

光

没有光,植物不能产生碳水化合物并将很快死去。植物每天光合作用的量部分取决于白天的长度和太阳光的强度。对于密集种植来说,明亮的阳光是最重要的。

白昼与黑夜周期的长度因与赤道的距离和季节变化而有所不同。这会影响作物开花结果,也是一个国家在引进新作物时的限制因素之一。我国正在实验,在这里白天的光照条件下,使草保持绿叶状态而不产生花芽。

水

水是所有植物细胞的重要组成部分,蒸腾作用也需要大量的水。水携带土壤中的营养成分进入植物,也把光合作用的产物从叶片运输到需要它们的地方。每产生1吨干物质植物需要吸收约200吨水。

二氧化碳(CO_2)

植物需要二氧化碳进行光合作用。二氧化碳通过气孔进入叶片,吸收的二氧化碳被蒸腾作用的速率影响。另一个限制因素是大气中二氧化碳的含量低(0.03%)。二氧化碳的百分比在富含有机质的土壤中会增加,因为其中的细菌活跃,能够释放二氧化碳。这可能是为什么在这样的土壤里作物生长更好的原因。

化学元素

植物为了建立自己的细胞结构,成为一个食品工厂,就需要许多简单的化学物质。这些物质从土壤溶液和黏土颗粒到根部。这些元素数额相当大——从几公斤到一百或几百公斤/公顷(每英亩需要几磅到一个百或几百磅重)被称为主要营养成分;那些需要少量的——几克到几公斤/公顷(一盎司分之几到几磅/公顷)是微量营养素和微量元素。

Unit 3

Plants Physiology

The seed does not fall on the soil but on the rubble. The living seed never sighs and feels pessimistic, because it is tempered by resistance.

——种子不落在肥土而落在瓦砾中,有生命力的种子决不会悲观和叹气,因为有了阻力才有磨炼。

Situational Dialogue:

An Order for Seeds

A: Welcome to our Seeds Business Hall. Can I help you?

B: I'd like to place an order for seeds with your company.

A: How much do you want?

B: About 10 tons.

A: What kind of varieties do you want?

B: What kind of hybridized rice developed by your company.

A: Is it the DJ-II type?

B: Yes. Can I enjoy a preferential price?

A: I can offer you a 5% reduction since it is our new product and I believe the prospect of market is promising.

B: But I still find it a little too high.

A: It all depends on the quality. That's the difference.

B: OK. That's accepted.

Notes

hybridize: 杂交

preferential: 特惠的

Lesson 1 Pollination

Pollination is the **transfer** of pollen from a **stamen** to a **pistil**. Pollination starts the production of seeds. Pollination is very important. It leads to the **creation** of new seeds that grow into new plants.

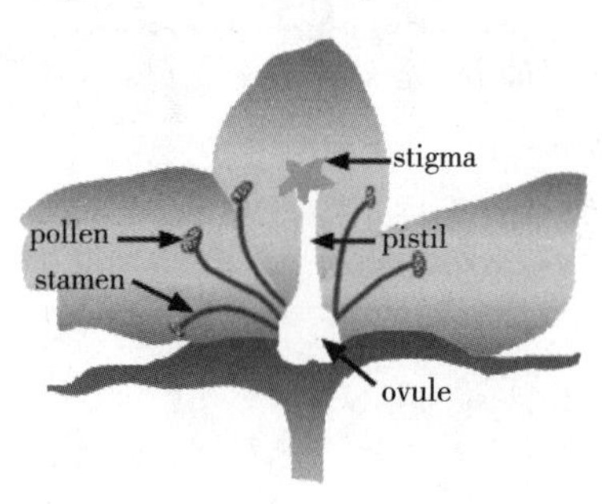

But how does pollination work? Well, it all begins in the flower. Flowering plants have several different parts that are important in pollination. Flowers have male parts called stamens that produce a **sticky powder** called pollen. Flowers also have a female part called the pistil. The top of the pistil is called the **stigma**, and is often sticky. Seeds are made at the base of the pistil, in the ovule.

To be pollinated, pollen must be moved from a stamen to the stigma. When pollen from a plant's stamen is transferred to that same plant's stigma, it is called **self-pollination**. When pollen from a plant's stamen is transferred to a different plant's stigma, it is called **cross-pollination**. Cross-pollination produces stronger plants. The plants must be of the same species. For example, only pollen from a daisy can pollinate another daisy. Pollen from a rose or an apple tree would not work.

But how does pollen from one plant get moved to another?

How Do Plants Get Pollinated?

Pollination occurs in several ways. People can transfer pollen from one flower to another, but most plants are pollinated without any help from people. Usually plants **rely on** animals or the wind to pollinate them.

When animals such as bees, **butterflies**, **moths**, **flies**, and **hummingbirds** pollinate plants, it's **accidental**. They are not trying to pollinate the plant. Usually they are at the plant to get food, the sticky pollen or a **sweet nectar** made at the base of the **petals**. When feeding, the animals accidentally **rub** against the stamens and get pollen **stuck** all over themselves. When they move to another flower to feed, some of the pollen can rub off onto this new plant's stigma.

Plants that are pollinated by animals often are **brightly colored** and have a strong

smell to attract the animal **pollinators**.

Another way plants are pollinated is by the wind. The wind picks up pollen from one plant and blows it onto another.

Plants that are pollinated by wind often have long stamens and pistils. Since they do not need to attract animal pollinators, they can be **dully colored**, **unscented**, and with small or no petals since no insect needs to land on them.

Some plants don't have flowers. Plants such as **mosses** and **ferns** reproduce by spores. **Cone-bearing** plants, like pine or **spruce** trees for example, reproduce by means of pollen that is produced by a male cone and travels by wind to a female cone of the same species. The seeds then develop in the female cone.

Vocabulary

transfer [træns'fɜː(r)] *v.* 使转移
stamen ['steɪmən] *n.* 雄蕊
pistil ['pɪstɪl] *n.* 雌蕊
creation [kri'eɪʃn] *n.* 制造,创造
sticky ['stɪki] *adj.* 黏性的
powder ['paʊdə(r)] *n.* 粉,粉末
stigma ['stɪgmə] *n.* [植]柱头
self-pollination ['selfˌpɒlɪ'neɪʃn] *n.* 自花授粉
cross-pollination [ˌkrɔːsˌpɒlɪ'neɪʃn] *n.* 异花授粉
rely on 信赖;依赖
butterfly ['bʌtəflaɪ] *n.* 蝴蝶
moth [mɒθ] *n.* 飞蛾,蛾子
fly [flaɪ] *n.* 苍蝇
hummingbird ['hʌmɪŋbɜːd] *n.* 蜂鸟
accidental [ˌæksɪ'dentl] *adj.* 意外的
sweet [swiːt] *n.* 糖果,甜食
nectar ['nektə(r)] *n.* 花蜜
petal ['petl] *n.* 花瓣
rub [rʌb] *v.* 擦,摩擦
stuck [stʌk] *adj.* 动不了的;被卡住的
brightly colored 颜色鲜艳的
pollinator ['pɒlɪneɪtə] *n.* 传粉者
dully colored 颜色沉闷的

unscented [ʌnˈsentɪd] *adj.* 无香味的
moss [mɒs] *n.* 苔藓
fern [ˈfɜːn] *n.* [植]羊齿植物,蕨类植物
cone-bearing [kəʊn ˈbeərɪŋ] *adj.* 带有球花的
spruce [spruːs] *n.* 针枞；云杉

Notes

1. Flowering plants have several different parts that are important in pollination. Flowers have male parts called stamens that produce a sticky powder called pollen.
   开花植物有几个传粉中重要的不同器官。花的雄性部分称为雄蕊,雄蕊产生黏性的粉末,称为花粉。
2. self-pollination：自花授粉。指一株植物的花粉,对同一个体的雌蕊进行授粉的现象。在两性花的植物中,又可分为同一花的雄蕊与雌蕊间进行授粉的同花授粉(菜豆属)和在一个花序(个体)中不同花间进行授粉的邻花授粉,以及同株不同花间进行授粉的同株异花授粉。被子植物大多为异花授粉,少数为自花授粉。高粱就是以自花授粉为主。
3. cross-pollination：异花授粉。异花授粉是植物界很普遍的授粉方式。它的花粉传播主要是靠风力或昆虫。依靠昆虫传粉的花叫作虫媒花,如梨、苹果、桃、柑橘、南瓜、油菜等植物的花。虫媒花的特点是花大,颜色鲜艳,有浓郁的香味和甜美的蜜汁,用于吸引昆虫,而且花粉较大,外壁有突起或粘质,很容易附着在昆虫的身体上。
4. People can transfer pollen from one flower to another, but most plants are pollinated without any help from people.
   人们可以把一朵花的花粉转移到另一朵花上,但大多数植物授粉是不需要任何人帮助的。
5. Usually they are at the plant to get food, the sticky pollen or a sweet nectar made at the base of the petals. When feeding, the animals accidentally rub against the stamens and get pollen stuck all over themselves.
   通常它们从植物中获取食物,这些食物可能是来自花瓣的黏性的花粉或甜的花蜜。舔食时,动物不小心蹭雄蕊,花粉就粘在它们身上。
6. Plants such as mosses and ferns reproduce by spore.
   如苔藓和蕨类植物,这些植物通过孢子繁殖。

Exercises

Ⅰ. Translate the following terms into Chinese.

1. pollination
2. pollen

3. stamen
4. pistil
5. seed
6. seedling
7. ovule
8. self-pollination
9. cross-pollination
10. sweet nectar
11. petal

Ⅱ. Fill in the blanks with the words given below. Change the forms where necessary.

attract	produce	male	start	creation
brightly	top	pollinate	rely	transfer

1. Pollination __________ the production of seeds.
2. It leads to the __________ of new seeds that grow into new plants.
3. Flowers have ____________ parts called stamens that produce a sticky powder called pollen.
4. The __________ of the pistil is called the stigma, and is often sticky.
5. Cross-pollination ________________ stronger plants.
6. People can __________ pollen from one flower to another, but most plants are pollinated without any help from people.
7. Usually plants __________ on animals or the wind to pollinate them.
8. When animals such as bees, butterflies, moths, flies, and hummingbirds pollinate plants, it's accidental. They are not trying to ______________ the plant.
9. Plants that are pollinated by animals often are ____________ colored and have a strong smell to attract the animal pollinators.
10. Since they do not need to ________ animal pollinators, they can be dully colored, unscented, and with small or no petals since no insect needs to land on them.

Ⅲ. Find out more information about pollinators.

Ⅳ. Read the text and answer the following questions.

1. What does the word "pollination" mean?

2. How does pollination work?

3. How many types of pollination do you know? What are they?

4. How do plants get pollinated?

5. What about plants that don't have flowers?

Learning for Fun

Sing a Pollination Song

How Do Plants Pollinate?

To the tune of: "This Land Is Your Land"

Play Verse 1

What does a plant need
To make a new seed?
Three things give flowers
Reproductive powers—
the sticky pollen,
the slender stamen,
and pistils make the flower whole.

Play Verse 2

What gets the pollen going
To keep new plants growing?
Different kinds of birds do,
Or the wind that's blowing.
Butterflies and bees,
Carry pollen they need
That's what makes pollination work.

Play Verse 3

If a flower's not scented,

Or brightly colored,
And the flowers are smaller
In clusters tighter
With stamens longer
the signs are stronger
This plant spreads pollen on the wind.

Play Verse 4

When bright colored flowers
Have a sweet perfume
And a sugary nectar
Then chances are good
That birds and insects active
Find the plants attractive
And they'll spread the pollen as they go.

参考译文

传　粉

传粉是花粉从雄蕊到雌蕊的转移。传粉始于种子的生产。授粉是非常重要的。它关系到能长成新生植物的新种子的产生。

但传粉是如何进行的？从花开始讲起。开花植物有几个传粉中重要的不同器官。花的雄性部分称为雄蕊,雄蕊产生黏性的粉末,称为花粉。花也有雌性的器官称为雌蕊。雌蕊的顶端被称为柱头,通常也是有黏性的。种子长在以雌蕊为基础的胚珠里。

为了传粉,花粉必须从雄蕊转移到柱头。当花粉从一个植物的雄蕊转移到同一植物的柱头,它被称为自花授粉。当花粉从一个植物的雄蕊转移到一个不同植物的柱头,它被称为异花授粉。异花授粉产生更强壮的植物。植物必须是同一物种。例如,一株雏菊只能由另一株雏菊的花粉粒授粉。一株玫瑰花或一棵苹果树的花粉是不行的。

但一株植物的花粉如何移动到另一株植物上呢？

植物如何被传粉？

传粉有几种方式。人们可以把一朵花的花粉转移到另一朵花上,但大多数植物授粉是不需要任何人帮助的。通常植物依靠动物或风授粉。

当动物如蜜蜂、蝴蝶、飞蛾、苍蝇和蜂鸟完成植物授粉时,是偶然的。它们不是有意识帮花授粉的。通常它们从植物中获取食物,这些食物可能是来自花瓣的黏性的花粉或甜的花蜜。舔食时,动物不小心摩擦雄蕊,花粉就粘在它们身上。当它们飞到另一朵花上采食时,一些花粉粒就会落到这朵花的柱头上。

动物传粉的植物往往是色彩鲜艳、具有吸引动物传粉者强烈气味的那些植物。

植物授粉的另一个方式是靠风。风吹起一个植株的花粉把它带到另一株植物上。

由风授粉的植物往往有很长的雄蕊和雌蕊。因为他们不需要吸引动物传粉者,它们可以是无色无味的,由于不需要昆虫在这些植物上驻足,所以这些植物的花瓣很小或者没有花瓣。

有些植物不开花。如苔藓和蕨类植物,这些植物通过孢子繁殖。带有球花的植物,如松树和云杉,雄球花产生的花粉通过风的吹动转移到同一个物种的雌球花中,它们以这种方式来繁殖。其种子在雌球花里生长。

Lesson 2 How Do Plants Get and Use Energy?

Leaves make glucose, which contains energy. Cells break down this energy. The plant uses the energy to live and grow.

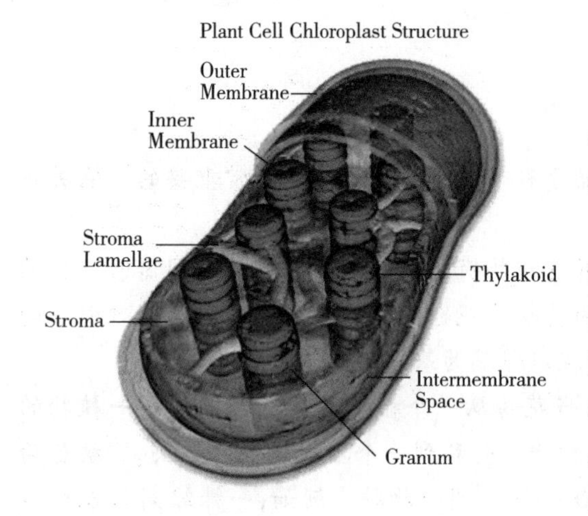

Photosynthesis

Different flowers have different colors, but most leaves are green at some stage in their life. Why are leaves green?

A **chloroplast** is a tiny structure found inside most plant cells. It contains a **substance** called **chlorophyll**, which makes leaves and other parts of plants green. Chlorophyll also slows a plant to make its own food in the form of glucose. Animal cells do not have chlorophyll, so they are usually not green and cannot make their own food.

Photosynthesis is the process in which plants make glucose. During photosynthesis, chlorophyll uses light energy from the sun, carbon dioxide from the air, and water to form glucose and **oxygen**. Energy is stored in glucose, which plants use for life processes. When **organisms** eat plants, they use glucose as energy too. The process of photosynthesis is **summarized** in this **equation**.

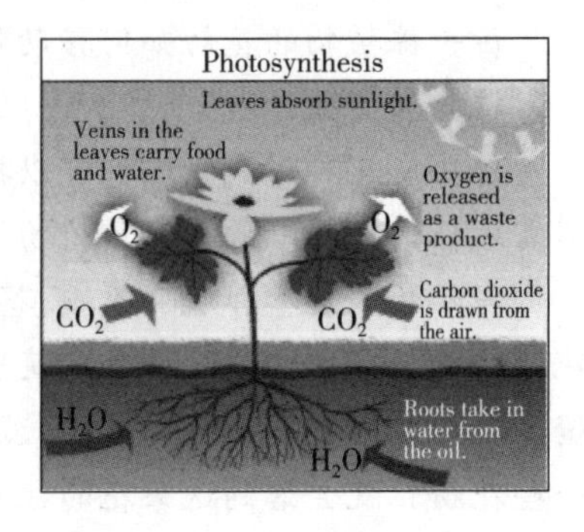

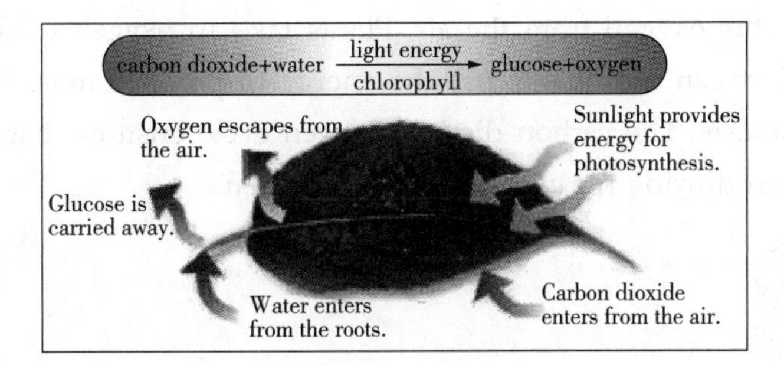

Transpiration

When stomata open, gases from the air enter the leaf and water passes out of the leaf. The loss of water from the leaf is called **transpiration**. The amount of water lost during transpiration depends on air **temperature**, wind, and the amount of water in the air and the soil.

To **survive**, the plant needs to **replace** the water lost during transpiration. As water **exits** a leaf, more water is taken into the leaf. This movement pulls more water up through the xylem in the stem. The plant can take in more water through its roots. If more water is lost by transpiration than is **gained** by the roots, the plant may **wilt** and even die.

Carbon Dioxide-Oxygen Cycle

Have you noticed that equations for photosynthesis and **cellular respiration** look **similar**? Let's look at the equations again.

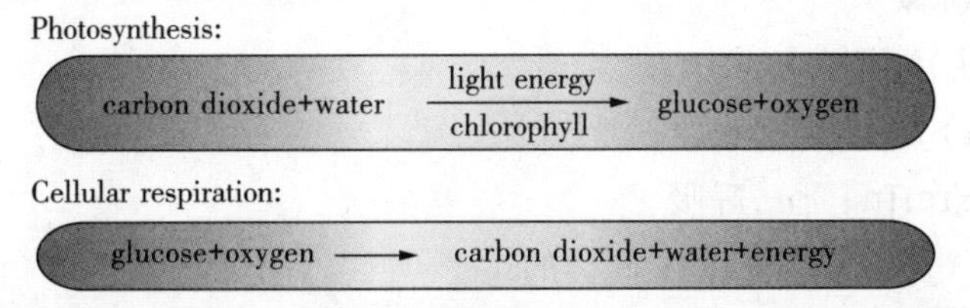

The two processes are almost the **reverse** of each other! Photosynthesis uses carbon dioxide, water, and energy to produce glucose and oxygen. Cellular respiration uses glucose and oxygen to produce carbon dioxide, water, and energy. If one process ended, the other process would not happen. Photosynthesis and cellular respiration form a **cycle**, called the carbon dioxide-oxygen cycle.

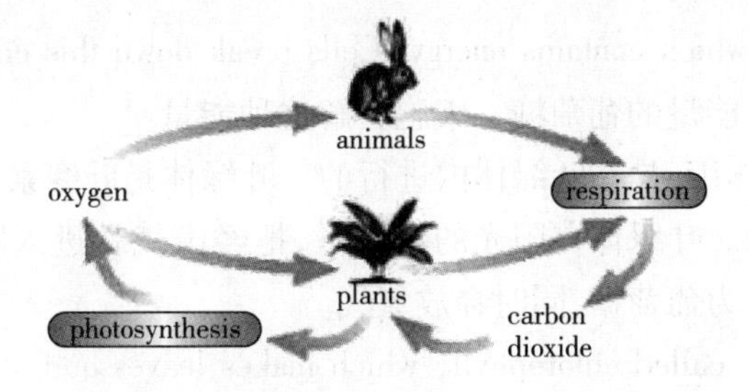

Animals breathe oxygen from the air. Plants take in oxygen to change energy in food to energy they can use. Plants use the energy to produce more food and oxygen during photosynthesis. The carbon dioxide-oxygen cycle ensures that there is enough oxygen and carbon dioxide for each process to happen.

Vocabulary

photosynthesis [ˌfəʊtəʊˈsɪnθəsɪs] *n.* 光合作用
chloroplast [ˈklɒrəplɑːst] *n.* 叶绿体
substance [ˈsʌbstəns] *n.* 物质,材料
chlorophyll [ˈklɒrəfɪl] *n.* 叶绿素
oxygen [ˈɒksɪdʒən] *n.* [化]氧,氧气
organism [ˈɔːgənɪzəm] *n.* 有机体
summarize [ˈsʌməraɪz] *v.* 总结,概述
equation [ɪˈkweɪʒn] *n.* 方程式
transpiration [ˌtrænspɪˈreɪʃn] *n.* 蒸发(物)
temperature [ˈtemprətʃə(r)] *n.* 温度;体温
survive [səˈvaɪv] *v.* 幸存
replace [rɪˈpleɪs] *vt.* 替换;代替
exit [ˈeksɪt] *v.* 离开;出去
gain [geɪn] *v.* 获得
wilt [wɪlt] *v.* (使)凋谢
cellular [ˈseljələ(r)] *adj.* 细胞的
respiration [ˌrespəˈreɪʃn] *n.* 呼吸
similar [ˈsɪmələ(r)] *adj.* 类似的;同类的
reverse [rɪˈvɜːs] *n.* 相反
cycle [ˈsaɪkl] *n.* 循环,周期

Notes

1. Leaves make glucose, which contains energy. Cells break down this energy.
植物的叶子生成包含能量的葡萄糖。细胞分解这种能量。
2. photosynthesis: 光合作用,是在叶绿体内进行的。叶绿体是叶绿素的载体。叶绿素是光合作用过程必不可少的。叶绿体在阳光的作用下,把经由气孔进入叶子内部的二氧化碳和由根部吸收的水转变为葡萄糖,同时释放氧气。
3. It contains a substance called chlorophyll, which makes leaves and other parts of plants green.

它含有一种叫叶绿素的物质,叶绿素使植物的叶子和其他部分呈现绿色。

葡萄糖是植物体内合成各种有机物的原料,而叶绿素则是植物叶子制造“粮食”的工厂。氮也是植物体内维生素和能量系统的组成部分。

4. transpiration:蒸腾作用,是水分从活的植物体表面(主要是叶子)以水蒸气状态散失到大气中的过程。与物理学的蒸发过程不同,蒸腾作用不仅受外界环境条件的影响,而且还受植物本身的调节和控制,因此它是一种复杂的生理过程。其主要过程为:土壤中的水分→根毛→根内导管→茎内导管→叶内导管→气孔→大气。植物幼小时,暴露在空气中的全部表面都能蒸腾。

5. When stomata open, gases from the air enter the leaf and water passes out of the leaf.

当气孔打开时,空气中的气体进入叶片,水从叶片中释放出来。

stoma:气孔,它是植物进行体内外气体交换的重要门户。水蒸气(H_2O)、二氧化碳(CO_2)、氧气(O_2)都要共用气孔这个通道,气孔的开闭会影响植物的蒸腾、光合、呼吸等生理过程。通过气孔的蒸腾,叫作气孔蒸腾,气孔蒸腾是植物蒸腾作用的最主要方式。

6. To survive, the plant needs to replace the water lost during transpiration.

为了生存,植物需要将蒸腾过程中丢失的水分补充回来。

survive *v.* ①生存; 存活; 幸存

the sequence of events that left the eight pupils battling to survive in icy seas for over four hours

致使这 8 个小学生在冰冷的海水里挣扎求生四个多小时的一连串事件

②挺过; 艰难度过

On my first day here I thought, “Oh, how will I survive?”

我第一天到这儿时就想,“天哪,我怎么能挺过去呢?”

③幸存; 幸免于难; 留存

When the market economy is introduced, many factories will not survive.

引入市场经济后,许多工厂将面临倒闭。

Exercises

I. Fill in the blanks with the words given below. Change the forms where necessary.

energy	summarize	survive	replace	exit
gain	similar	reverse	ensure	happen

1. The question is whether the government can pass laws to __________ employees' benefits.
2. Few birds managed to ________ the bad winter.
3. ____________ things have happened elsewhere.
4. His conversation brims with __________.
5. It was clear: we should ____________.
6. When will your cellphone ____________ your wallet?

7. What the IBM's ________ is clear enough.

8. How do you __________ Google's culture?

9. Mary, did something ______ today at work?

10. This does not ________ that trend.

Ⅱ. Translate the following sentences into Chinese.

1. A chloroplast is a tiny structure found inside most plant cells. It contains a substance called chlorophyll, which makes leaves and other parts of plants green.

2. During photosynthesis, chlorophyll uses light energy from the sun, carbon dioxide from the air, and water to form glucose and oxygen.

3. The amount of water lost during transpiration depends on air temperature, wind, and the amount of water in the air and the soil.

4. If more water is lost by transpiration than is gained by the roots, the plant may wilt and even die.

5. Photosynthesis uses carbon dioxide, water, and energy to produce glucose and oxygen. Cellular respiration uses glucose and oxygen to produce carbon dioxide, water, and energy.

6. Animals breathe oxygen from the air. Plants take in oxygen to change energy in food to energy they can use. Plants use the energy to produce more food and oxygen during photosynthesis. The carbon dioxide-oxygen cycle ensures that there is enough oxygen and carbon dioxide for each process to happen.

参考译文

植物如何获取和使用能量？

植物的叶子生成包含能量的葡萄糖。细胞分解这种能量。植物利用这些能量存活、生长。

光合作用

不同的花有不同的颜色，但大多数的叶子在其生长的某个阶段都是绿色的。为什么叶子是绿色的？

叶绿体是大多数植物细胞中的一个微小结构。它含有一种叫叶绿素的物质，叶绿素使植物的叶子和其他部分呈现绿色。叶绿素也减缓了植物以葡萄糖的形式生成自身食物的速度。动物细胞没有叶绿素，所以他们通常不是绿色的，也不能为自己生产食物。

光合作用是植物生成葡萄糖的过程。光合作用中，叶绿素利用来自太阳光、空气中的二氧化碳和水形成葡萄糖和氧气。能量储存在葡萄糖里，植物运用这些葡萄糖维持生命。当生物体吃下这种植物时，他们也把葡萄糖作为了能量。光合作用可以概括为这个方程。

$$CO_2+H_2O \xrightarrow[\text{叶绿素}]{\text{光}} (CH_2O)+O_2$$ [(CH_2O)表示糖类；$C_6H_{12}O_6$为葡萄糖]

蒸腾作用

当气孔打开时，空气中的气体进入叶片，水从叶片中释放出来。从叶片里损失水称为蒸腾。在蒸腾作用中，损失的水量取决于空气的温度、风、空气中的含水量和土壤。

为了生存，植物需要将蒸腾过程中丢失的水分补充回来。叶子里损失了水，叶子就需要吸收更多的水。这一过程从茎的木质部运送更多的水。植物可以通过它的根吸收更多的水。如果蒸腾作用时损失的水比根吸收的水更多，植物就会枯萎甚至死亡。

碳氧循环

你有没有注意到，光合作用和细胞呼吸方程类似？让我们再看一次两个方程。

光合作用：$CO_2+H_2O \xrightarrow[\text{叶绿素}]{\text{光}} (CH_2O)+O_2$ [(CH_2O)表示糖类；$C_6H_{12}O_6$为葡萄糖]

蒸腾作用：$(CH_2O)+O_2 \longrightarrow CO_2+H_2O+E$

这两个过程几乎完全相反！光合作用以二氧化碳、水和能量生成葡萄糖和氧气。细胞呼吸则用葡萄糖和氧气生成二氧化碳、水和能量。如果一个进程结束，其他过程也不会发生。光合作用和细胞呼吸作用形成一个循环，称为二氧化碳—氧气循环。

动物吸入空气中的氧气。为了使能量为其所用，植物吸收氧气改变食物中的能量。在光合作用过程中植物利用能量生产更多的食物和氧气。二氧化碳—氧气循环确保每一个过程都有足够的氧气和二氧化碳。

Lesson 3 Making Food

Plants are very important to us. All food people eat comes **directly** or indirectly from plants.

Directly from Plants:	Indirectly from Plants:
For example, apples come from an apple tree. The **flour** used to make bread comes from a **wheat** plant.	**Steak** comes from a cow, and we all know that cows are animals, not plants, right? But what does the cow eat? It eats grass and grains—plants!

So all the foods we eat come from plants. But what do plants eat? They make their own food!

What Do Plants Need to Make Food?

Plants need several things to make their own food.

They need:

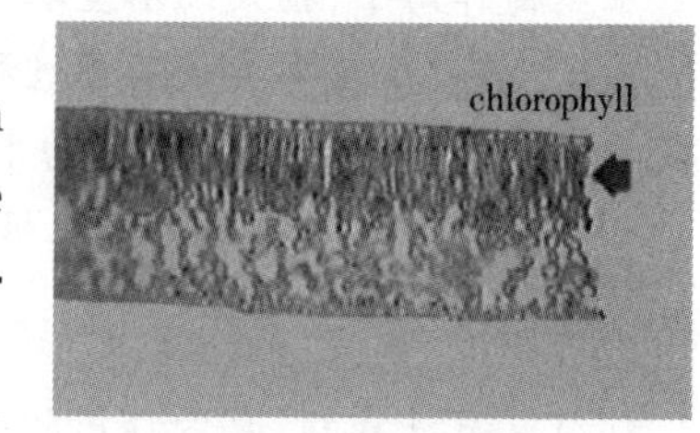

- **chlorophyll**, a green **pigment** or color, found in the leaves of plants that helps the plant make food (see the layer of chlorophyll in the cross-section of a leaf at right);
- light (either natural sunlight or **artificial** light, like from a **light bulb**);
- carbon dioxide (CO_2)(a gas found in the air; one of the gases people and animals **breathe out** when they **exhale**);
- water (which the plant **collects** through its roots);

- nutrients and minerals (which the plant collects from the soil through its roots).

Plants make food in their leaves. The leaves contain a pigment called chlorophyll, which colors the leaves green. Chlorophyll can make food the plant can use from carbon dioxide, water, nutrients, and energy from sunlight. This

process is called photosynthesis.

During the process of photosynthesis, plants release oxygen into the air. People and animals need oxygen to breathe.

Seed Dispersal

People plant some seeds, but most plants don't rely on people. Plants rely on animals and wind and water to help **scatter** their seeds.

- Animal Dispersal

Animals **disperse** seeds in several ways. First, some plants, like the **burr** at right, have **barbs** or other structures that get **tangled** in animal **fur** or **feathers**, and are then carried to new sites. Other plants produce their seeds inside **fleshy** fruits that then get eaten by an animal. The fruit is **digested** by the animal, but the seeds pass through the digestive **tract**, and are dropped in other locations. Some animals bury seeds, like **squirrels** with **acorns**, to save for later, but may not return to get the seed. It can grow into a new plant.

- Wind Dispersal

The kind of seeds which are often wind dispersed are smaller seeds that have **wings** or other hair-like or feather-like structures. Plants that produce wind blown seeds, like **dandelions**, often produce lots of seeds to ensure that some of the seeds are blown to areas where the seeds can germinate.

- Floating in Water

Many **aquatic** plants and plants that live near water have seeds that can float, and are carried by water. Plants living along **streams** and rivers have seeds that float downstream, and therefore become germinate at new sites. The size of the seed is not a **factor** in determining whether or not a seed can float. Some very large seeds, like **coconuts**, can float. Some small seeds also float.

And some plants disperse their seeds in other ways. Some plants have **unique** ways to disperse their seeds. Several kinds of plants "**shoot**" seeds out of **pods**. The seeds can travel quite a few feet from the plant this way.

Vocabulary

directly [dəˈrektli] *adv.* 直接地
flour [ˈflaʊə(r)] *n.* 面粉
wheat [wiːt] *n.* 小麦
steak [steɪk] *n.* 牛排；肉排
chlorophyll [ˈklɒrəfɪl] *n.* 叶绿素
pigment [ˈpɪgmənt] *n.* [生]色素
artificial [ˌɑːtɪˈfɪʃl] *adj.* 人造的；人工的
light bulb 灯泡
breathe out 呼气；呼出
exhale [eksˈheɪl] *v.* 发散出；放出
collect [kəˈlekt] *v.* 收集
dispersal [dɪˈspɜːsl] *n.* 散布；分散
scatter [ˈskætə(r)] *v.* 分散
disperse [dɪˈspɜːs] *v.* (使)分散
burr [bɜː(r)] *n.* 带刺的种子
barb [bɑːb] *n.* [植]钩状毛
tangle [ˈtæŋgl] *v.* (使)缠结
fur [fɜː(r)] *n.* 毛皮；软毛
feather [ˈfeðə(r)] *n.* 羽毛
fleshy [ˈfleʃi] *adj.* (似)肉的；多肉的
digest [daɪˈdʒest] *v.* 消化；整理
tract [trækt] *n.* (神经纤维的)束
squirrel [ˈskwɪrəl] *n.* 松鼠
acorn [ˈeɪkɔːn] *n.* 橡子
wing [wɪŋ] *n.* 翅膀，翼
dandelion [ˈdændɪlaɪən] *n.* [植]蒲公英
float [ˈfləʊt] *v.* (使)浮动；(使)漂浮
aquatic [əˈkwætɪk] *adj.* 水生的
stream [striːm] *n.* 河流，小河
factor [ˈfæktə(r)] *n.* 因素
coconut [ˈkəʊkənʌt] *n.* [植]椰子
unique [juˈniːk] *adj.* 唯一的，仅有的
shoot [ʃuːt] *v.* 射出
pod [pɒd] *n.* 荚，豆荚

Notes

1. Chlorophyll, a green pigment or color, found in the leaves of plants that helps the plant make food (see the layer of chlorophyll in the cross-section of a leaf at right).
   叶绿素,是一种绿色的颜料或颜色,它存在于植物的叶子中,能够帮助植物制造食物(见右图叶片截面中的叶绿素层)。
2. First, some plants, like the burr at right, have barbs or other structures that get tangled in animal fur or feathers, and are then carried to new sites.
   首先,一些植物,像右图的毛刺,有可以缠在动物毛皮或羽毛上的倒钩或其他结构,可以把种子带到新的地方。
3. Other plants produce their seeds inside fleshy fruits that then get eaten by an animal.
   其他植物种子长在有肉质的果实里,这些果实被动物吃下。
4. Some animals bury seeds, like squirrels with acorns, to save for later, but may not return to get the seed. It can grow into a new plant.
   有些动物为了储存以后的食物把种子埋起来,像松鼠埋橡子一样,但可能不会再回来取这些储存的种子。

Exercises

Ⅰ. Read the text and decide whether each of the following statements is True (T) or False (F).

1. All food people eat comes directly or indirectly from plants.
2. Plants make their own food using water, chlorophyll, light nutrients and minerals and O_2!
3. CO_2 is a gas found in the air; one of the gases people and animals breathe out when they exhale.
4. Chlorophyll can make food the plant can use from carbon dioxide, water, nutrients, and energy from sunlight.
5. People plant some seeds, but most plants rely on people. Plants rely on animals and wind and water to help scatter their seeds.
6. The kind of seeds which are often animal dispersed are smaller seeds that have wings or other hair-like or feather-like structures.
7. The size of the seed is a factor in determining whether or not a seed can float.

Ⅱ. Read the text and answer the following questions.

1. What do plants need to make food?

2. What is photosynthesis?

3. How does seed disperse? What does the word "dispersal" mean?

Ⅲ. Translate the following sentences.

1. During the process of photosynthesis, plants release oxygen into the air. People and animals need oxygen to breathe.

2. First, some plants, like the burr at right, have barbs or other structures that get tangled in animal fur or feathers, and are then carried to new sites.

3. The fruit is digested by the animal, but the seeds pass through the digestive tract, and are dropped in other locations.

4. Plants that produce wind blown seeds, like dandelions, often produce lots of seeds to ensure that some of the seeds are blown to areas where the seeds can germinate.

5. Many aquatic plants and plants that live near water have seeds that can float, and are carried by water.

参考译文

提供食物

植物对我们很重要。人吃的所有食物都直接或间接来自植物。

<table>
<tr><td>直接来自植物：</td><td>间接来自植物：</td></tr>
<tr><td> 例如，苹果来自苹果树。做面包用的面粉来源于小麦。</td><td> 牛排来自牛，我们都知道牛是动物而非植物，对吧？但是牛吃什么呢？它吃草和谷物——这些是植物！</td></tr>
</table>

所以我们所吃的所有食物都来自植物。但是植物吃什么呢？它们自己制造食物！

植物需要什么制造食物？

植物需要几种物质才能制造出自己的食物。

它们需要：

- 叶绿素，是一种绿色的颜料或颜色，它存在于植物的叶子中，能够帮助植物制造食物（见右图的叶片截面中的叶绿素层）；
- 光（自然光或人工光，如灯光）；
- 二氧化碳（CO_2）（是一种空气中的气体；也是人和动物呼吸时呼出的气体之一）；
- 水（植物通过根收集而来）；
- 营养素和矿物质（植物通过根从土壤里吸收而来）。

植物在叶子中制造食物。叶子中含有被称为叶绿素的色素，叶绿素使叶子变绿。叶绿素可以运用二氧化碳、水、营养和来自太阳的光能为植物制造食物。这一过程叫作光合作用。

光合作用过程中，植物释放氧气到空气中。人和动物都需要氧气来呼吸。

- **种子传播**

人们播撒一些种子，但大多数植物不依靠人。植物依靠动物和风还有水帮助它们播撒种子。

- **动物传播**

动物传播种子有几种方式。首先，一些植物，像右图的毛刺，有可以缠在动物毛皮或羽毛上的倒钩或其他结构，可以把种子带到新的地方。其他植物的种子长在有肉质的果实里，这些果实被动物吃下。水果被动物消化，但种子则经过消化道散落在其他地方。有些动物为了储存以后的食物把种子埋起来，像松鼠埋橡子一样，但可能不会再回来取这些储存的种子。种子就会长成一棵新的植物。

- **风传播**

由风传播的种子通常都是小的，有翼或其他毛发或羽毛状结构的种子。种子通过风吹动传播的植物，像蒲公英，往往会产生大量的种子，以确保种子可以吹到能令其发芽的地方。

- **浮在水里**

许多水生植物和生活在水边的植物结出可以漂浮并能被水携带的种子。生活在溪水和河流边的植物种子漂浮到水的下游,因此在新的地方萌发。种子的大小不是决定种子是否漂浮的因素。一些像椰子一样大的种子可以漂浮,一些小的种子也能漂浮。

有些植物通过其他的方式传播种子。有些植物以非常独特的方式来播撒它们的种子。有几种植物的种子是从种子荚中“射出”的。种子可以以这种方式旅行好几英尺。

Unit 4
Food

> *Food is nature. Chew quietly and savor gently. It's extraordinary.*
> ——食为天性，静静地咀嚼，轻轻地回味，非比寻常的韵致。

Situational Dialogue:

Order for Peanut Meat

A: Good morning.

B: Good morning. I am interested in your peanut meat. Could you help me with that?

A: Certainly. May I know how much you would like to order?

B: More than 10 tons of Shandong peanut meat if the price is competitive.

A: Wait a moment, please. Let me have a check. The goods you want to order are on the 9th floor. I think you would like to have a look, wouldn't you? This way, please.

B: OK! After you.

A: Here we are. These are what you want.

B: Thank you. The peanuts are OK, but I wonder whether you can make us a more favorable offer.

A: As your order is quite a big one, I may think it worthwhile to make a concession.

B: Fine, how about the packing?

A: In polybags, two bags per carton, 25 kg net.

B: Then let's come down to the discounts you are going to allow.

A: OK. How about...

polybag：塑料袋
carton：纸箱

Lesson 1 Agriculture, Farmers, and the Modern Food Industry

Modern food is a **miracle** of sorts. You can eat a fruit from **halfway** around the world, and it tastes nearly as fresh as if it had just been picked. **Agriculture** and farmers are changing rapidly. Once **primarily** a poor man's vocation, farming has become big business, and small farms are gradually disappearing, being replaced by agricultural **enterprises**. Consumers should learn about this shift in modern farming practices and how it affects you.

How Has Agriculture Changed?

Modern farmers use technological advancements and agricultural **innovations** to improve their harvests and reduce the impact of **insects** and other **pests** on crops. The use of drip irrigation, **fertilizers**, both natural and chemical, and even **infrared** or radar surveys of soil have helped to cut losses and improve yields drastically. Health problems and reduced soil quality have been linked to the use of some modern technologies for the farm, however.

How Have Farmers Responded to These Changes?

The image of a farmer that most people think of is a self-employed homeowner with a large property used for farming. This doesn't match modern farmers, however. Many farmers work for corporate farming enterprises and have college degrees or even graduate education in agriculture or similar fields. Farming is a complex modern enterprise, and farmers have had to change with the times. Small farms have largely disappeared, as agricultural enterprises buy larger and larger plots of land. Some small farms do persist, but their operators are often very well-educated and interested in their field or consult with individuals who have specialized education in farming, in order to improve their farm's output. Subsistence farming was once the norm, but it is quickly becoming the exception.

Is Modern Agriculture All about Business?

Despite the influence of business on modern agriculture, small farms are still crucial to communities. Many small towns and cities now have programs for community supported agriculture, which allow participants to receive produce directly from a small farmer. These individuals, as well as consumers who grow their own produce, often care very little about the business end of modern agriculture. Another interesting trend is the farm to table movement in cuisine, which has led to many restaurants and grocery stores to grow their own produce.

What Farmers Plant?

In addition to modifying how plants are watered and fertilized, modern agriculture has affected the decision of what to plant. Farmers can **assess** the **acidity** of soil to determine what crops will thrive best in a particular field. Additionally, the careful study of plant growth and behavior has allowed for a greater understanding of companion plant relationships. Some plants help each other to grow and thrive, and farmers are using this knowledge to improve plant health and harvest yields. Race crops seeds, carefully selected seeds coming from each year's harvests, and planted the following year for the cultivation of specific characteristics, are decreasing in popularity. Industrial seed stocks are taking the place of these seeds.

Modern farming is very different from farming 50 years ago. Machines, technology, and modern innovations have changed the agricultural industry drastically. Basic farming practices and the education level of the farmer has changed significantly. Farming is now big business.

Vocabulary

miracle [ˈmɪrəkl] *n.* 奇迹
halfway [hɑːfˈweɪ] *adv.* 在中途
agriculture [ˈægrɪkʌltʃə] *n.* 农业
primarily [ˈpraɪmerəli] *adv.* 主要地，根本上
enterprise [ˈentəpraɪz] *n.* [经] 企业
innovation [ɪnəˈveɪʃn] *n.* 创新，革新
insect [ˈɪnsekt] *n.* 昆虫
pest [pest] *n.* 害虫
fertilizer [ˈfɜːtɪlaɪzə] *n.* [肥料] 肥料
infrared [ɪnfrəˈred] *n.* 红外线

assess [əˈses] *v.* 评定；估价
acidity [əˈsɪdəti] *n.* 酸度

Notes

1. The image of a farmer that most people think of is a self-employed homeowner with a large property used for farming.
   至于农民的形象，大多数人认为是一个把大量财产用于农业的自营业主。
   think of：认为　　self-employed：个体经营的
2. Race crops seeds, carefully selected seeds coming from each year's harvests, and planted the following year for the cultivation of specific characteristics, are decreasing in popularity.
   作物种子，都是从每一年的收成中精心挑选的，为了培育作物栽培的某些特点，需要第二年继续种植，但这种方式已在大量减少。
   主语是 race crops seeds；谓语是 are decreasing。selected seeds… and planted…中 selected 和 planted，过去分词作定语，表示被动含义。

Exercises

Ⅰ. Fill in the blanks with the words given below. Change the forms where necessary.

self-employed	enterprises	image	match	agriculture

The ________ of a farmer that most people think of is a ________ homeowner with a large property used for farming. This doesn't ________ modern farmers, however. Many work for corporate farming ________ and have college degrees or even graduate education in ________ or similar fields.

Ⅱ. Choose the best answer from the choices given below.

1. In the end, he __________.
   A. got invited　　B. gets invited　　C. was invited　　D. was to be invited
2. He wore dark glasses to avoid __________.
   A. having been spotted　　B. to be spotted　　C. spotted　　D. being spotted
3. A new theory __________ before it can be put into practice.
   A. must be tested　　B. be tested　　C. can be tested　　D. to be tested
4. Sooner than __________ for others, she started her own business.
   A. working　　B. worked　　C. to work　　D. work
5. I can't afford as __________ car as this one.
   A. expensive a　　B. an expensive
   C. a more expensive　　D. a most expensive

Ⅲ. Translate the first paragraph of the text into Chinese.

参考译文

农业,农民以及现代食品工业

现代食品是一项奇迹。你可以吃到来自世界各地的水果,新鲜得就如刚刚摘下来一样。农业和农民正在迅速改变。农业曾经一度是穷人的职业,现在已成为大生意。小型农场正逐渐消失,被农业企业所代替。消费者应该了解这种现代农业实践的转变以及它如何影响你。

农业发生了怎样的变化?

现代农民使用先进技术和农业创新以增加他们的收成,并减少昆虫和其他害虫对农作物的影响。使用滴灌、天然肥料或化肥,甚至用红外或雷达测量土壤,有助于减少损失和大幅度提高产量。然而,健康问题和降低土壤质量已经和农场的一些现代技术的使用相关联。

农民如何应对这些变化?

至于农民的形象,大多数人认为就是把大量财产用于农业的自营业主。但这不符合现代农民的现状。许多农民为农业企业工作,他们拥有农业或相关领域的大学学历甚至研究生教育背景。农业是一个复杂的现代行业,农民必须紧跟时代的变化。小农场已经大量消失,因为农业企业购买了大量的土地。一些小农场能够坚持做下去,但他们的经营者往往受过良好的教育,并对他们的领域非常感兴趣,为了提高农业产出,咨询农业领域的专家。自给农业曾经是典范,但它正在成为例外。

现代农业都是商业吗?

尽管商业对现代农业有影响,但小农场对社区来讲仍然至关重要。许多小城镇和城市现在有社区支持农业项目,它允许参与者直接从小农场收农产品。这些个体和自己生产农产品的消费者的农民,自己生产,很少关心现代农业的商业端。另一个有趣的趋势是美食从农场到餐桌的变化,这就促使许多餐馆和杂货店自己种植和生产。

农民种什么?

除了改进如何浇水与如何施肥,现代农业也影响着农民种植什么。农民可以判断土壤酸度确定什么地方种什么长得最好。此外,仔细研究植物的生长特点可以更好地了解混栽植物之间的关系。有些植物能有助于彼此茁壮成长,农民们利用这些知识来改善植物健康,提高粮食产量。作物种子,是从每一年的收成中精心挑选的,为了培育作物的某些特点,需要第二年继续种植,但这种方式正在大量减少。所以工业化生产的种子正在代替这种种子。

现代农业与50年前截然不同。机器、技术和现代创新极大地改变了农业产业。基本耕作模式和农民的受教育水平有了明显的改变。农业现在是大产业。

Lesson 2 Foods That Help You Stay Hydrated

According to the old rule, you're supposed to drink eight glasses of water per day. That can seem like a **daunting** task on some days, but here's the catch: You don't have to drink all that water. Roughly 20% of our daily H_2O intake comes from solid foods, especially fruits and vegetables.

It's still important to drink plenty of water—especially in the summertime—but you can also **quench** your thirst with these hugely hydrating foods, all of which are at least 90% water by weight.

Cucumber

Water content: 96.7%

This summer **veggie**—which has the highest water content of any solid food—is perfect in salads.

Want to **pump** up cucumber's hydrating power even more? Try blending it with **nonfat yogurt**, **mint**, and ice cubes to make cucumber soup. "Soup is always hydrating, but you may not want to eat something hot in the summertime," Gans says. "Chilled cucumber soup, on the other hand, is so refreshing and delicious any time of year."

Celery

Water content: 95.4%

That urban legend about **celery** having negative calories isn't quite true, but it's pretty close. Like all foods that are high in water, celery has very few calories—just 6 calories per stalk. And its fiber and water helps to fill you up and **curb** your appetite.

This lightweight veggie isn't short on nutrition, however. Celery contains folate and vitamins A, C, and K. And thanks in part to its high water content, celery neutralizes stomach **acid** and is often recommended as anatural remedy for heartburn and acid reflux.

Radishes

Water content: 95.3%

These refreshing root vegetables should be a fixture in your spring and summer

salads. They provide a burst of spicy-sweet flavor—and color! —in a small package, and more importantly they're filled with antioxidants such as catechin.

Tomatoes

Water content: 94.5%

"Sliced and diced tomatoes will always be a mainstay of salads, sauces, and sandwiches, but don't forget about sweet cherry and grape varieties, which make an excellent hydrating snack," Gans says. "They're great to just pop in your mouth, maybe with some nuts or some low-sodium cheese," she says. "You get this great explosion of flavor when you bite into them."

Cauliflower

Water content: 92.1%

Don't let **cauliflower**'s pale complexion fool you: In addition to having lots of water, these unassuming florets are packed with vitamins and phytonutrients that have been shown to help lower cholesterol and fight cancer, including breast cancer. (A 2012 study of breast cancer patients by Vanderbilt University researchers found that eating cruciferous veggies like cauliflower was associated with a lower risk of dying from the disease or seeing a recurrence.)

"Break them up and add them to a salad for a satisfying crunch," Gans suggests. "You can even skip the croutons!"

Watermelon

Water content: 91.5%

It's fairly obvious that watermelon is full of, well, water, but this juicy melon is also among the richest sources of lycopene, a cancer-fighting antioxidant found in red fruits and vegetables. In fact, watermelon contains more lycopene than raw tomatoes—about 12 milligrams per wedge, versus 3 milligrams per medium-sized tomato.

Although this melon is plenty hydrating on its own, Gans loves to mix it with water in the summertime. "Keep a water pitcher in the fridge with watermelon cubes in the bottom," she says. "It's really refreshing, and great incentive to drink more water overall."

Spinach

Water content: 91.4%

Iceberg lettuce may have a higher water content, but spinach is usually a better bet overall. Piling raw spinach leaves on your sandwich or salad provides nearly as much built-in hydration, with an added nutritional punch.

Spinach is rich in lutein, **potassium**, fiber, and brain-boosting **folate**, and just one cup of raw leaves contains 15% of your daily intake of vitamin E—an important antioxidant for fighting off the damaging molecules known as free radicals.

Strawberries

Water content: 91.0%

All berries are good foods for hydration, but juicy red strawberries are easily the best of the bunch. Raspberries and blueberries both hover around 85% water, while blackberries are only slightly better at 88.2%.

"I love strawberries blended in a smoothie or mixed with plain nonfat yogurt—another hydrating food," Gans says. Strawberries add natural sweetness to the yogurt, she adds, and the combo of carbohydrates, fiber, and protein make a great post-workout recovery snack.

Broccoli

Water content: 90.7%

Like its cousin cauliflower, raw **broccoli** adds a satisfying crunch to a salad. But its nutritional profile—lots of fiber, potassium, vitamin A, and vitamin C—is slightly more impressive. What's more, broccoli is the only cruciferous vegetable (a category that contains cabbage and kale, in addition to cauliflower) with a significant amount of sulforaphane, a potent compound that boosts the body's protective **enzymes** and flushes out cancer-causing chemicals.

Vocabulary

hydrate [ˈhaɪdreɪt] *v.* (使)成水化合物
daunting [ˈdɔːntɪŋ] *adj.* 使人畏缩的
quench [kwentʃ] *v.* 熄灭；平息
cucumber [ˈkjuːkʌmbə] *n.* 黄瓜
veggie [ˈvedʒi] *n.* 素食者
pump [pʌmp] *v.* 打气
nonfat [ˈnɒnˈfæt] *adj.* 脱脂的
yogurt [ˈjɒgət] *n.* 酸奶酪,[食品] 酸乳
mint [mɪnt] *n.* 薄荷
celery [ˈseləri] *n.* 芹菜
curb [kɜːb] *v.* 控制
acid [ˈæsɪd] *n.* 酸
reflux [ˈriːflʌks] *n.* 逆流

cauliflower [ˈkɒlɪflaʊə] *n.* 花椰菜,菜花
potassium [pəˈtæsɪəm] *n.* [化学] 钾
folate [ˈfəʊleɪt] *n.* 叶酸
broccoli [ˈbrɒkəli] *n.* 花椰菜; 西兰花
enzyme [ˈenzaɪm] *n.* [生化] 酶

Notes

1. According to the old rule, you're supposed to drink eight glasses of water per day.
   根据过去的经验法则,你应该每天喝八杯水。
   rule of thumb: 经验法则; be supposed to: 应该,被期望
2. (But) here's the catch: You don't have to drink all that water.
   但必须注意的一点是,你没必要全喝水。
3. Piling raw spinach leaves on your sandwich or salad provides nearly as much built-in hydration, with an added nutritional punch.
   堆在三明治或沙拉上的生菠菜叶提供了几乎同样多的水分,此外还有营养。

Exercises

Ⅰ. Fill in the blanks with the words given below. Change the forms where necessary.

yogurt	summertime	cucumber	hydrating	refreshing	chill

Want to pump up cucumber's ____________ power even more? Try blending it with nonfat ________, mint, and ice cubes to make __________ soup. "Soup is always hydrating, but you may not want to eat something hot in the __________," Gans says. "________ cucumber soup, on the other hand, is so ________ and delicious any time of year."

Ⅱ. Read the text and answer the following questions.

1. Can you list the name of the high-water-content foods?

2. What is the characteristic of the celery?

3. What's the benefit of the cauliflower for your health?

Ⅲ. Translate the following sentences into Chinese.

1. According to the old rule, you're supposed to drink eight glasses of water per day.

2. It's still important to drink plenty of water—especially in the summertime.

3. Raspberries and blueberries both hover around 85% water, while blackberries are only slightly better at 88.2%.

4. Spinach is rich in lutein, potassium, fiber, and brain-boosting folate, and just one cup of raw leaves contains 15% of your daily intake of vitamin E—an important antioxidant for fighting off the damaging molecules known as free radicals.

参考译文

帮你保持水分的食物

根据过去的经验法则，你应该每天喝八杯水。有时候这似乎是一项艰巨的任务，但必须注意的一点是，你没必要全喝水，大约20%的水来自我们日常摄入的固体食物，特别是水果和蔬菜。

喝大量的水很重要，尤其是在夏天，但你也可以通过吃这些含水量大的食物来解渴，所有这些食物至少含90%的水分。

黄　瓜

含水率：96.7%

这种夏季蔬菜（所有固体食物中水分含量最高）是做沙拉的完美食材。

要想提高黄瓜的保湿力？尝试将其与脱脂酸奶、薄荷和冰块一块做成黄瓜汤。“汤总是补水，但是在夏季你可能不想吃热的，”甘斯说。“换个做法，黄瓜冷汤，在任何时候都是那么的清爽美味。”

芹　菜

含水率：95.4%

传说芹菜不含热量,这个说法不完全正确,但很接近了。像所有含水分高的食物一样,芹菜的热量很低,每棵菜只有6卡路里。其高纤维和水分让你有饱腹感,因此可以控制你的食欲。

这种不起眼的蔬菜并不缺乏营养,相反,芹菜富含叶酸和维生素A、C、K,部分是由于其水分含量高,芹菜中和胃酸,经常被作为一种治疗烧心和反酸的天然药品。

萝 卜

含水率:95.3%

这些令人耳目一新的块根类蔬菜应该是你春夏季沙拉的常用食品。它们又甜又辣,色彩鲜艳,更重要的是它们富含抗氧化剂如儿茶素。

西红柿

含水率:94.5%

"切碎的西红柿是沙拉、酱料和三明治的主要原料,但是别忘了甜樱桃和各种葡萄,这些都是极好的保湿小吃。"甘斯说:"它们刚刚放进你的嘴里的时候感觉非常好,也可以加一些坚果或一些低钠的奶酪。"她说,"当你咬的时候,浓郁的味道会弥漫开来。"

菜 花

含水率:92.1%

不要让菜花苍白的颜色欺骗你:除了含水量高,这个不起眼的菜还富含维生素和植物营养素,已证明菜花有助于降低胆固醇和对抗癌症,包括乳腺癌。(2012年,研究乳腺癌患者的范德堡大学的人员发现,吃十字花科蔬菜比如花椰菜,能降低疾病的死亡率和复发率。)

"把菜花切开加到你的沙拉上,"甘斯建议,"你甚至可以不要油炸面包丁!"

西 瓜

含水率:91.5%

西瓜含水量高这很明显,但这种多汁的瓜也含有丰富的番茄红素,一种在红色水果和蔬菜中发现的对抗癌症的抗氧化剂。事实上,西瓜的番茄红素含量比生西红柿还高。每块约12毫克,而一个中等大的番茄只含3毫克。

尽管西瓜本身含大量的水,在夏天甘斯却喜欢把它和水搅拌在一起。"把一个底下放着西瓜块的水壶放在冰箱里,"她说,"这真的很提神,和水一块喝下去很刺激。"

菠 菜

含水率:91.4%

生菜通常有较高的水含量,但菠菜是更明智的选择。堆在三明治或沙拉上的生菠菜叶提供了几乎同样多的水分,此外还有营养。

菠菜中含有丰富的叶黄素、纤维、钾,还有促进大脑的叶酸,仅仅一杯生菠菜叶就含有每日摄入的维生素E(一种抵抗有害自由基分子的重要的抗氧化剂)的15%。

草 莓

含水率:91%

所有浆果都是保湿性很好的食物,但多汁的红草莓是最好的。树莓和蓝莓含水量都徘徊在85%,而黑莓略高,有88.2%。

"我喜欢把草莓放在鲜果奶昔里或是放在脱脂酸奶中,就是另一种保湿的食物,"甘斯说。草莓加入酸奶,等于加入了天然的甜味,她补充说,结合了碳水化合物和蛋白质、纤维,是一种训练后恢复体力的极好的小吃。

西兰花

含水率:90.7%

像它的近亲菜花,生西兰花可以做成可口的沙拉。其营养价值丰富,富含纤维、钾、维生素A、维生素C。更重要的是,西兰花是唯一的含有大量萝卜硫素的十字花科蔬菜(一种包括卷心菜、甘蓝,还有菜花的蔬菜),萝卜硫素是一种有效的化合物,能增强身体的保护酶活性,清除致癌化学物质。

Lesson 3 Tomato Sauce

What Is Tomato Sauce?

Tomato **sauce** is a kind of **condiment** served along with the food to enhance its taste and flavor. Because of rich flavor and **tempting** sweet and sour **aroma**, tomato sauce has become a popular accompaniment for most of the dishes, especially **pastas**, cooked in households. It is semi-liquid in content and tempered with salt. Sometimes, few drops of olive oil are also added in its seasoning.

Tomato sauce is an all-time kid's favorite. Grown-ups also cannot resist the taste of tomato sauce for its sweet and sour flavor. It is meant for every age and **goes well with** every taste.

Whether it's a small party with friends or a large family gathering, you can self cook the **tangy** tomato sauce or get it from market to add to the richness of the dishes.

Uses of Tomato Sauce

Who doesn't enjoy that tangy taste, taking over other flavors in mouth completely to melt and merge with them, and to leave us **crave** for more? You can use it as topping on your dish or can create a base with it to make your **recipe** richer and delicious. Tomato sauce is used as base in all the pasta dishes. However, it adds to the flavors of cooked chicken, fish **fillets**, lamb **casseroles** and pork items when applied as topping.

Types of Tomato Sauce

You can use in your cooking homemade tomato sauce or the ready made one. Going by taste, you would find purely tomato, onion, **garlic**, red **chilly** or **soya bean** flavor in tomato sauce. There are numerous other varieties also available in the market.

Tomato sauce variety can also be defined in terms of the methods of preparation. You can **incorporate** uncooked tomato sauce in your dish to add a **refreshing** aroma to it. You have half-or semi-cooked tomato sauce for the dishes that require light **soothing** smell. Lastly, you have well simmered tomato sauce. It is thick and has a mix of sweet and sour flavor.

Tomato Sauce Preparations

You can use as many **ingredients** as you want to cook tomato sauce. However, if you want an all-time cooking solution, you are recommended to prepare your sauce with some basic ingredients such as olive oil, peeled garlic, tomatoes and salt. The cooking time will be 4 hours, approximately. You can add other flavors later on to this basic content when required. Once the sauce is ready, you can use it in any Thai or Italian cuisines among others.

Sauce Quality & Safety—Quality Assurance

Most of the key brand companies as well as small companies need to test their product content before launch. The quality control analysts take care of the examining part of the product from the **initial** stage itself. The raw materials for the tomato sauce are checked for their quality. Companies have to be careful throughout the processing, packaging and delivery methods. The food analysts need to check the following factors:

- Use of ingredients;
- Quality of flavors and edible;
- Level of toxicity;
- Nutrition level;
- Level of impurities.

Sauce Quality & Safety—Hygiene & Packaging

While launching tomato sauce in the market, the companies need to pay attention to **hygiene** and packaging. The products need to be sterilized. There should be a vacuum pack to ensure safety of the product. The freshness and purity of the products need to be maintained. When the product is being packaged, the following guidelines need to be followed by the manufacturers:

- Correct manufacturing date;
- Correct expiry date;
- Labeling of the used ingredients;
- Quantity.

A can of tomato paste

Tinned mackerel fillet in tomato sauce is a popular food in Scandinavia.

Vocabulary

sauce [sɔːs] *n.* 沙司；调味汁
condiment [ˈkɒndɪmənt] *n.* 调味品；佐料
tempting [ˈtemptɪŋ] *adj.* 吸引人的；诱惑人的
aroma [əˈrəʊmə] *n.* 芳香
pasta [ˈpæstə] *n.* 意大利面食
go well with 协调；和……很相配
tangy [ˈtæŋi] *adj.* 扑鼻的
crave [kreɪv] *v.* 渴望
recipe [ˈresəpi] *n.* 食谱
fillet [ˈfɪlɪt] *n.* [食品]肉片
casserole [ˈkæsərəʊl] *n.* 勺皿
garlic [ˈgɑːlɪk] *n.* 大蒜
chilly [ˈtʃɪli] *n.* 辣椒
soya [ˈsɔɪə] *n.* 大豆,[作物] 黄豆
bean [biːn] *n.* 豆
incorporate [ɪnˈkɔːpəreɪt] *v.* 包含,吸收
refreshing [rɪˈfreʃɪŋ] *adj.* 提神的；使清爽的

soothing [ˈsuːðɪŋ] *adj.* 抚慰的；使人宽心的
ingredient [ɪnˈgriːdɪənt] *n.* 材料；原料
assurance [əˈʃʊərəns] *n.* 保证，担保
initial [ɪˈnɪʃl] *adj.* 最初的
hygiene [ˈhaɪdʒiːn] *n.* 卫生

Notes

1. Who doesn't enjoy that tangy taste, taking over other flavors in mouth completely to melt and merge with them, and to leave us crave for more?
   谁不喜欢那种扑鼻的香味呢，在口中完全代替了其他味道，慢慢融化，并与其他味道融合，让我们更加渴望得到它。
   take over：掌握，接管，接手；crave for：渴望。
2. However, if you want an all-time cooking solution, you are recommended to prepare your sauce with some basic ingredients such as olive oil, peeled garlic, tomatoes and salt.
   如果你想要制作传统的番茄酱，建议你准备一些基本材料，例如橄榄油、剥皮大蒜、西红柿和盐。
   all-time：历来的
   Pen and paper—One of my all-time favorites.
   纸和笔——我一直以来最推崇的工具之一。

Exercises

Ⅰ. Fill in the blanks with the words given below. Change the forms where necessary.

freshness	ensure	package	guideline	purity

There should be a vacuum pack to __________ safety of the product. The __________ and _________ of the products need to be maintained. When the product is being _________, the following _________ need to be followed by the manufacturers.

Ⅱ. Read the text and answer the following questions.

1. What is tomato sauce?

2. What are the main types of tomato sauce?

3. What are the basic ingredients you prepare when you make the tomato sauce?

Ⅲ. Translate the following sentences into Chinese.

1. Because of rich flavor and tempting sweet and sour aroma, tomato sauce has become a popular accompaniment for most of the dishes, especially pastas, cooked in households.

2. Tomato sauce is an all-time kid's favorite. Grown-ups also cannot resist the taste of tomato sauce for its sweet and sour flavor.

3. You can use in your cooking homemade tomato sauce or the ready made one.

4. Most of the key brand companies as well as small companies need to test their product content before launch.

5. While launching tomato sauce in the market, the companies need to pay attention to hygiene and packaging.

参考译文

番茄酱

什么是番茄酱?

番茄酱是一种调味料,用来增强食物的口感和风味。由于它浓郁的味道和诱人的酸甜味,对于大多数菜肴来说,成了必不可少的一味调料,尤其是在家烹饪意大利面的时候。番茄酱是加了盐的半流质,有时候,为了增加它的风味,会加几滴橄榄油。

番茄酱历来都是孩子们的最爱。成年人也无法抗拒番茄酱这种酸甜的风味,这意味着番茄酱老少咸宜,它也可以和任何口味搭配。

无论是与友人小聚还是家庭大聚会,你都可以自制或是从市场购买番茄酱来增加菜肴的口味。

番茄酱的使用

谁不喜欢那种扑鼻的香味呢,在口中完全代替了其他味道,慢慢融化,并与其他味道融合,让我们更加渴望得到它。你可以用它点缀你的菜,或是成为菜的基本组成部分,来丰富你的菜肴,增加美味度。番茄酱在所有的意大利面食中是必不可少的,而在烹饪鸡肉、鱼片、羊肉砂锅和猪肉的时候则作为点缀。

番茄酱的种类

你可以使用自制的番茄酱,也可以使用现成的。番茄酱有各种风味的,纯番茄味的、洋葱味的、大蒜味的、红辣椒或是大豆味的,在超市还有各种口味的番茄酱。

番茄酱种类也可以按照制作方法分类。有增加菜肴清香味的生番茄酱,有较淡的半加工的番茄酱,还有用文火慢炖的熟番茄酱,它有味道浓厚的酸甜味。

制作番茄酱的准备工作

制作番茄酱需要准备许多材料。如果你想要制作传统的番茄酱,建议你准备一些基本材料,例如橄榄油、剥皮大蒜、西红柿和盐。制作时间大约4个小时,如需要可以添加其他的调料。制成后,就可以用在任何泰国菜或意大利美食上了。

番茄酱质量安全——质量保证

大多数品牌公司和中小企业在产品上市之前都会对产品做测试。质量控制分析师从最初阶段开始就会关注、检查产品,首先是原材料的检查,在整个加工、包装、送货过程中,公司都会非常小心。食品分析师需要检查以下要素:

- 使用的原料
- 风味和食用品质
- 毒副作用
- 营养
- 杂质

番茄酱质量安全——卫生与包装

在市场上推出番茄酱的同时,企业需要注意卫生和包装。产品必须消毒。应该用真空包装以确保产品的安全性。产品的新鲜度和纯度需要维护。当产品被包装后,制造商必须遵循下列指导方针:

- 正确的生产日期
- 正确的截止日期
- 使用成分标签
- 数量

Lesson 4 Celery—an Ancient Healing Food

Celery has a long history of use and is truly an ancient **healing** food. At first glance celery may seem rather unimpressive, but the more you look into its **background** and **medicinal** uses, the more you realize that we must have been misinformed on the **array** of **ailments** and was very bitter in taste. It is believed to have originated from the Mediterranean basin, and has been harvested since about 850 BC. Its medicinal **properties** are beleved to be from its **volatile** oils which are found in all parts of the plant, but seem to be **concentrated** in its seeds. **Physicians** used celery to treat colds, flu, water **retention**, poor digestion, **arthritis**, **liver**, and **spleen** ailments.

Our common celery **stalk** is mostly composed of about 83% water and a healthy amount of **fiber**. This is well known. What is not well known is the fact that celery also contains many **micronutrients** which it receives from rain, sunshine, and the soil **medium** from which it is grown. We have been led to believe that celery does not have much to it when in fact celery truly contains a wealth of health improving nutrients that we can obtain from it. Celery has a **profile** that is much more than water and fiber!

The effects of celery on the body are **diuretic**, **expectorant**, **carminative**, anti-asthmatic, and digestive aid. Celery is a good source of vitamin K, vitamin C, **potassium**, **folate**, and fiber. It also contains **molybdenum**, **manganese**, vitamin B1, vitamin B2, vitamin B6, **calcium**, magnesium, **phosphorus**, **iron**, and **tryptophan**. Celery also contains about 35 **milligrams** of a beneficial **sodium complex**; this is useful in reducing stomach acid levels and raising our **hydrochloric** acid leels which in turn improves digestion.

Celery has been used for many years in Chinese medicine to alleviate high blood pressure. It is believed that the phthalides in celery relax the arteries and allow the **vessels** to **dilate** which enables the blood to flow more freely. These phthalides also relieve our atress hormones and in turn the less stressed our body is the lower our blood pressure become. Celery is also a very good source of potassium, calcium and **magnesium**, all of which have been associated with reduced blood pressure.

Celery is also a good source of vitamin C, and along with that comes all the benefits that vitamin C carries with it. Some of these benefits include a **boost** in the **immune** system and a reduction in the **symptoms** and severity of cold. Vitamin C is also an **antioxidant** which has been shown to lower **inflammation** in many cases such as arthritis and **asthma**. Vitamin C is crucial in the production of **collagen**.

Vocabulary

celery [ˈseləri] *n.* 芹菜
heal [hiːl] *v.* 使恢复正常
background [ˈbækgraʊnd] *n.* 背景
medicinal [məˈdɪsɪnl] *adj.* 医学的；药用的
array [əˈreɪ] *n.* 阵列
ailment [ˈeɪlmənt] *n.* 疾病(尤指微恙)
property [ˈprɒpəti] *n.* 特性,属性
volatile [ˈvɒlətaɪl] *adj.* 易变的
concentrate [ˈkɒnsntreɪt] *v.* 集中
physician [fɪˈzɪʃn] *n.* 医生,内科医生
retention [rɪˈtenʃn] *n.* 保留；记忆力,保持力；滞留
arthritis [ɑːˈθraɪtɪs] *n.* 关节炎
liver [ˈlɪvə(r)] *n.* 肝脏
spleen [spliːn] *n.* 脾
stalk [stɔːk] *n.* [植]茎
fiber [ˈfaɪbə] *n.* 纤维
micronutrient [ˌmaɪkrəʊˈnjuːtriənt] *n.* 微量营养素
medium [ˈmiːdiəm] *n.* 媒介物
profile [ˈprəʊfaɪl] *n.* 轮廓
diuretic [ˌdaɪjuˈretɪk] *n.* [医]利尿剂
expectorant [ɪkˈspektərənt] *n.* 除痰剂
carminative [ˈkɑːmɪnətɪv] *adj.* 排出胃肠气体的

potassium [pəˈtæsiəm] *n.* [化]钾
folate [ˈfəʊleɪt] *n.* 叶酸
molybdenum [məˈlɪbdənəm] *n.* 钼
manganese [ˈmæŋgəniːz] *n.* [化]锰
calcium [ˈkælsiəm] *n.* [化]钙
phosphorus [ˈfɒsfərəs] *n.* [化]磷
iron [ˈaɪən] *n.* 铁
tryptophan [ˈtrɪptəfæn] *n.* [生化]色氨酸
milligram [ˈmɪligræm] *n.* 毫克(千分之一克)
sodium [ˈsəʊdiəm] *n.* [化]钠
complex [ˈkɒmpleks] *n.* 建筑群；相关联的一组事物
hydrochloric [ˈhaɪdrəˈklɒrɪk] *adj.* 含氢和氯的
vessel [ˈvesl] *n.* 血管
dilate [daɪˈleɪt] *v.* 使扩大；使膨胀
magnesium [mægˈniːziəm] *n.* [化]镁
boost [buːst] *n.* 提高,增加；帮助
immune [ɪˈmjuːn] *adj.* 免疫的；有免疫力的
symptom [ˈsɪmptəm] *n.* 症状
antioxidant [ˌæntiˈɒksɪdənt] *n.* 抗氧化剂,硬化防止剂
inflammation [ˌɪnfləˈmeɪʃn] *n.* [医]炎症
asthma [ˈæsmə] *n.* [医]气喘,哮喘
collagen [ˈkɒlədʒən] *n.* [生化]胶原蛋白

Notes

1. at first glance：一看就,初看,乍看起来
2. originate from：源于……
3. the Mediterranean：地中海
4. be composed of：由……组成
5. anti-asthmatic：平喘的
6. sodium complex：钠复合
7. Chinese medicine：中药

Exercises

Ⅰ. Put the following terms into English.

1. 芹菜
2. 纤维
3. 茎
4. 胶原蛋白
5. 色氨酸
6. 抗氧化剂
7. 主茎
8. 叶酸
9. 微量营养素
10. 钙

Ⅱ. Put the following terms into Chinese.

1. arthritis
2. ailment
3. medicinal
4. physician
5. liver
6. inflammation
7. asthma
8. symptom
9. anti-asthmatic
10. vessel

Ⅲ. Translate the following sentences into English.

1. 乍一看,芹菜好似很普通,但是对它的背景和疗效了解越多,就越会意识到我们对它的使用产生的误解有多深。

2. 普通芹菜的主茎通常由约83%的水和适量的纤维组成。

3. 由于研究天然食品的方法不断发展,我们对饭桌上吃的东西越来越了解,正如我们刚刚发现的,芹菜确实不只是水和纤维。

4. 芹菜还是维生素C的良好来源,包括维生素C带来的所有益处。

Ⅳ. Translate the following paragraphs into Chinese.

Safety concerns over the drinking of raw milk and consuming products made from raw milk have been around for many decades. Both the proponents and opponents of raw milk consumption continue to make a case for their side of the issue, and both sides have valid points.

Those acquainted in any way at all with agriculture in general, and dairying in particular, know the risks associated with the consumption of raw milk. It's certainly no secret that milk, when not properly handled, can harbor dangerous pathogens that can quickly and seriously sicken us. However, many continue to ask for and consume raw milk. Ironically, "buyer beware" is a great concept until it happens to us.

In spite of the risks, for many, the consumption of raw milk and its prductsrepresents a basic tenant of healthy eating and consumer rights. Their lifestyles and values include consuming foods that are "natural" and free from real or perceived potential adulterations—those contaminations or compromises coming from modern commercial agriculture. Even though the pasteurization of milk is the accepted norm for the vast majority of the dairy industry, there's an increasing percentage of the population that are willing to accept the potential risks that are inherent with the consumption of raw milk.

V. Read the text again and discuss the following question with your classmates.

As we all know that celery is truly much more than just water and fiber, can you name some of its medicinal uses in the history?

参考译文

芹菜——古老的疗疾祛病食物

芹菜的使用有着漫长的历史，确切地说，它是古老的疗疾祛病食物。乍一看，芹菜好似很普通，但是对它的背景和疗效了解越多，就越会意识到我们对它的使用产生的误解有多深。传统上，芹菜用来治疗一系列小病，而且味苦。据悉它来自地中海盆地，约在公元前850年就有收获。其医药性质据悉来自自身挥发的油，蕴藏于植物所有部分，不过好似在种子部分更为集中。医师用芹菜治疗感冒、流感、闭尿、消化不良、关节炎、肝脾疾病。

众所周知，普通芹菜的主茎通常由约83%的水和适量的纤维组成。可大家有所不知的是，芹菜还含有许多微量营养物，来自孕育植物生长的雨水、阳光和土壤环境。我们一直被引导认为芹菜并没有多少这些物质，可实际上，它包含着丰富的、有益健康的、我们可吸取的营养素。芹菜的营养远远不只水和纤维!

芹菜对身体的疗效有利尿、化痰、排胃气、平喘及助消化。它是维生素K、维生素C、钾、叶酸和纤维的良好来源。还含有钼、锰、维生素B1、维生素B2、维生素B6、钙、镁、磷、铁和色氨酸。它还含有约35毫克的有益身体的钠合成物,对降低胃酸程度及提升盐酸水准有作用,因而能相应地改善消化。

在中医里芹菜一直用于降高血压。据悉芹菜所含的苯酞物质可以放松动脉,扩张血管,使血液更自由地流动。这种苯酞物质还可以减轻压力荷尔蒙,使身体减压,从而降低血压。芹菜还是钾、钙、镁的良好来源,所有这些营养都和降低血压有关。

芹菜还是维生素C的良好来源,具有维生素C带来的所有益处。包括提升免疫力以及降低感冒的症状和严重程度。维生素C还是一种抗氧化剂,可以减缓许多病症,如关节炎和哮喘的炎症症状。维生素C对产生胶原蛋白起着关键作用。

Unit 5
Agricultural Technology

Science and technology are productive forces, and also the primary productive forces.

——科学技术是生产力，而且是第一生产力。

Situational Dialogue:

Visiting a Flower Farm

A: I'm going to give you a brief account of my flower farm. We grow chrysanthemum here. We have four large greenhouses and two small ones.

B: Would you please explain how to grow seedlings?

A: Of course, I will. We choose the pieces of chrysanthemum-leaves and put the root-inducing powder at their ends and then transplant them into the flowerbeds in the small greenhouse.

B: The flowers may need certain temperature and humidity. How do you keep that?

A: We have a large compressor to regulate temperature and an automatic water-supply system to control the humidity.

B: Thank you. What about the market of the flowers?

A: They sell very well in both the domestic and international market.

B: How do you transport them?

A: In containers.

B: Thank you very much for the information.

A: You're welcome.

chrysanthemum：菊花
seedling：秧苗
the root-inducing powder：生根粉
humidity：湿度
compressor：空气压缩机
container：集装箱

Lesson 1 Smart Agriculture

Climate change, population growth and increasingly **scarce** resources are putting agriculture under **pressure**. Farmers must harvest as much as possible from the smallest possible land surface. Until now, the industry **confronted** this **challenge** with **innovations** in individual sectors: Intelligent systems regulate engines in order to save on gas, for instance. With the **aid** of **satellites** and sensor technology, farming equipment can automatically **perform** the field work; in doing so, they efficiently **distribute** seed, fertilizer and pesticides on the arable land. **Nonetheless**, **optimization** is **gradually** hitting its limits.

Intelligent Networking as Competitive Advantage

For their **exhibit**, the experts **intentionally** chose the field of agriculture: A **miniature** tractor with an **implement** moves across a plot of land on an agricultural **diorama**. Located at the edge of the **farmland** are two **tablet** PCs. Visitors to the trade show can use them to start up the automated control of the farm equipment. Six **screens** are **suspended** above the **model** farm. They **display** the processes behind the automation, showing how **software** manages the **functionality**. Today's tractors and implements feature **extensive** use of **electronics** and software—these are known as "embedded systems". The **motto** of the exhibit is "SEE: Software Engineering Explained". The **visualization** helps visitors to understand the challenges and **solutions** of **interconnecting** embedded systems and IT systems.

The networking of agricultural operations is not limited to simple task management for agricultural **machinery**. Over the last few years, the number of "players" in agricultural business has **soared**: Besides seed and fertilizer producers, sensor technology and data service providers are joining in the mix, offering **geodata** and weather data, for instance; systems for e-government and smartphone **apps** for identifying pests are also used. "The challenge lies in linking all systems intelligently, and in creating standards for **interfaces** so that all **participants** can benefit," says Dr. Jens Knodel, Smart Farming project manager.

Vocabulary

climate [ˈklaɪmət] *n.* 气候
scarce [skeəs] *adj.* 缺乏的
pressure [ˈpreʃə(r)] *n.* 压力
confront [kənˈfrʌnt] *v.* 面对；使面对面，使对质
challenge [ˈtʃæləndʒ] *n.* 挑战
innovation [ˌɪnəˈveɪʃn] *n.* 改革，创新
aid [eɪd] *n.* 帮助
satellite [ˈsætəlaɪt] *n.* 卫星；人造卫星
perform [pəˈfɔːm] *v.* 执行；履行
distribute [dɪˈstrɪbjuːt] *v.* 分配，散布
nonetheless [ˌnʌnðəˈles] *adv.* 虽然如此
optimization [ˌɒptɪmaɪˈzeɪʃn] *n.* 最佳化，最优化
gradually [ˈgrædʒuəli] *adv.* 逐步地，渐渐地
competitive [kəmˈpetətɪv] *adj.* 竞争的
exhibit [ɪgˈzɪbɪt] *n.* 展览，陈列
intentionally [ɪnˈtenʃənəli] *adv.* 有意地
miniature [ˈmɪnətʃə(r)] *adj.* 小型的
implement [ˈɪmplɪmənt] *n.* 工具，器械
diorama [ˌdaɪəˈrɑːmə] *n.* 透视画，立体模型
farmland [ˈfɑːmlænd] *n.* 农田
tablet [ˈtæblət] *n.* 碑，匾
screen [skriːn] *n.* 屏幕；银幕
suspend [səˈspend] *v.* 暂停；悬挂
model [ˈmɒdl] *n.* 模型
display [dɪˈspleɪ] *v.* 显示；陈列
software [ˈsɒftweə(r)] *n.* 软件
functionality [ˌfʌŋkʃəˈnæləti] *n.* 功能；功能性

extensive [ɪk'stensɪv] *adj.* 广阔的，广大的
electronics [ɪˌlek'trɒnɪks] *n.* 电子学
motto ['mɒtəʊ] *n.* 座右铭；格言
visualization [ˌvɪʒʊəlaɪ'zeɪʃn] *n.* 形象化，想象
solution [sə'luːʃn] *n.* 解决
interconnect [ˌɪntəkə'nekt] *v.* 使互相连接
machinery [mə'ʃiːnəri] *n.* （总称）机器
soar [sɔː(r)] *v.* 高飞越过
geodata [dʒiːəʊ'deɪtə] *n.* 地理数据
app [æp] *n.* 计算机应用程序
interface ['ɪntəfeɪs] *n.* 界面
participant [pɑː'tɪsɪpənt] *n.* 参加者，参与者

Notes

1. smart agriculture：智能农业（或称工厂化农业），是指在相对可控的环境条件下，采用工业化生产，实现集约高效可持续发展的现代超前农业生产方式，就是农业先进设施与露地相配套、具有高度的技术规范和高效益的集约化规模经营的生产方式。它集科研、生产、加工、销售于一体，实现周年性、全天候、反季节的企业化规模生产；它集成现代生物技术、农业工程、农用新材料等学科，以现代化农业设施为依托，科技含量高，产品附加值高，土地产出率高和劳动生产率高，是我国农业新技术革命的跨世纪工程。智能农业产品通过实时采集温室内温度、土壤温度、CO_2浓度、湿度信号以及光照、叶面湿度、露点温度等环境参数，自动开启或者关闭指定设备。可以根据用户需求，随时进行处理，为设施农业综合生态信息自动监测、对环境进行自动控制和智能化管理提供科学依据。通过模块采集温度传感器等信号，经由无线信号收发模块传输数据，实现对大棚温湿度的远程控制。智能农业还包括智能粮库系统，该系统通过将粮库内温湿度变化的感知与计算机或手机的连接进行实时观察，记录现场情况以保证量粮库的温湿度平衡。
2. Until now, the industry confronted this challenge with innovations in individual sectors: Intelligent systems regulate engines in order to save on gas, for instance.
   目前为止，该行业面临着各个方面的创新与挑战：例如，智能系统为了节省汽油调节发动机。
3. With the aid of satellites and sensor technology, farming equipment can automatically perform the field work; in doing so, they efficiently distribute seed, fertilizer and pesticides on the arable land.
   借助卫星和传感器技术的帮助，农业设备可以自动执行现场工作；这样的话，它们在耕地上有效地播撒种子、化肥和农药。
4. For their exhibit, the experts intentionally chose the field of agriculture: A miniature tractor

with an implement moves across a plot of land on an agricultural diorama.

他们的展示里，专家们有意选择了农业领域：在一个农业的立体模型上，一台带有工具的微型拖拉机在一块地上移动。

5. Today's tractors and implements feature extensive use of electronics and software—these are known as "embedded systems". The motto of the exhibit is "SEE: Software Engineering Explained".

今天使用的拖拉机和工具以广泛应用电子设备和软件为特色——这些被称为"嵌入式系统"。展会的座右铭是"看：软件工程的解释"。

6. The visualization helps visitors to understand the challenges and solutions of interconnecting embedded systems and IT systems.

可视化可以帮助游客了解互连嵌入式系统和 IT 系统的挑战和解决方案。

7. The networking of agricultural operations is not limited to simple task management for agricultural machinery.

农业操作网络不只限于农业机械的简单的任务管理。

Ⅰ. Fill in the blanks with the words given below. Change the forms where necessary.

manage	agriculture	innovation	challenge	resource
benefit	process	distribute	confront	process

1. Climate change, population growth and increasingly scarce ________ are putting agriculture under pressure.
2. Until now, the industry ________ this challenge with innovations in individual sectors: Intelligent systems regulate engines in order to save on gas, for instance.
3. With the aid of satellites and sensor technology, farming equipment can automatically perform the field work; in doing so, they efficiently ________ seed, fertilizer and pesticides on the arable land.
4. They map the entire ________ electronically, from the farm computer to the harvesting operation.
5. Visitors to the trade show can use them to start up the automated ________ of the farm equipment.
6. They display the processes behind the automation, showing how software ________ the functionality.
7. The visualization helps visitors to understand the ________ and solutions of interconnecting embedded systems and IT systems.
8. Indeed, it is becoming the next major factor in ________ in several industries.

9. The challenge lies in linking all systems intelligently, and in creating standards for interfaces so that all participants can __________, says Dr. Jens Knodel, Smart Farming project manager.
10. "It is not restricted to __________, but may be of interest to small and medium-sized enterprises, for instance," says Knodel.

Ⅱ. Translate the following paragraph into Chinese.

For their exhibit, the experts intentionally chose the field of agriculture: A miniature tractor with an implement moves across a plot of land on an agricultural diorama. Located at the edge of the farmland are two tablet PCs. Visitors to the trade show can use them to start up the automated control of the farm equipment. Six screens are suspended above the model farm. They display the processes behind the automation, showing how software manages the functionality. Today's tractors and implements feature extensive use of electronics and software—these are known as "embedded systems". The motto of the exhibit is "SEE: Software Engineering Explained". The visualization helps visitors to understand the challenges and solutions of interconnecting embedded systems and IT systems.

Ⅲ. Supplemental reading.

What is the main idea of the passage? How do you understand this question?

New vs. Old

Plant scientists have been able to overcome the slow reproductive phase involved in growing plants and have been able to produce plants that can grow in half the time with much better results.

Although old methods of plant breeding are still practiced today in many parts of the world, especially in developing countries, many developed countries have switched to using these new technologies. However, as for anything there are limitations and concerns for both methods.

The more primitive (原始的) methods of selection take a long time and do not always provide the desired crops. In addition, when a farmer does finally achieve his desired plant product, industry has changed and a new trait is preferred over the previous one. In a constantly growing and changing world, it seems that the farmer can never get ahead. Also, changing environmental conditions can lead to decreased crop yield from year to year and possibly even cause the farmer to lose entire crops.

In the more modern methods of plant growth, crops can be grown in artificial medium requiring less landmass (土地) to produce larger amounts of crops in less time. Although this seems like a great alternative to the earlier methods, it can also be devastating. By growing plants at a faster rate, we may be losing the essential vitamins and nutrients that are important for us.

Incorporating foreign (外来的) genes has been suggested to lead to allergies (过敏) in many people as well as affect the food chain of other organisms.

Plant science is a growing field that is constantly evolving to promote changes that benefit both the industry and accommodate (供给) the growing populations that we are facing.

参考译文

智能农业

气候变化,人口增长和资源日益稀缺,使农业承受了沉重的压力。农民必须从最少的土地中取得尽可能多的收获。目前为止,该行业面临着各个部分的创新与挑战:例如,智能系统为了节省汽油调节发动机。借助卫星和传感器技术的帮助,农业设备可以自动执行现场工作;这样的话,它们在耕地上有效地在土地上播撒种子、施肥和洒农药。然而,它的优势逐渐暴露出了其限制。

以智能网络为竞争优势

他们的展示里,专家们有意选择了农业领域:在一个农业的立体模型上,一台带有工具的微型拖拉机在一块地上移动。农田边缘有两台平板电脑。展会与会者可以用它们来启动农场设备的自动化控制。农场模型上方悬挂着六个屏幕。他们显示自动化的过程,展示软件如何管理各项功能。今天使用的拖拉机和工具以广泛应用电子设备和软件为特色——这些被称为“嵌入式系统”。展会的座右铭是“看:软件工程的解释”。可视化可以帮助游客了解互连嵌入式系统和 IT 系统的挑战和解决方案。

农业操作网络不只限于农业机械的简单的任务管理。在过去的几年中,农业中“玩家”的数量大幅增加:除了种子和化肥生产商,传感器技术和数据服务提供商都加入进来了,例如,提供地理和气象数据;电子政务系统和用于识别害虫的智能手机应用程序。“挑战在于智能化地连接所有系统,并建立让所有参与者受益的标准的接口。”智能农业项目经理 Jens Knodel 博士说。

Lesson 2 Automated Agriculture

"We can use **automation** to increase efficiency and yield, by having many of the manual tasks of farming performed by specially designed agricultural robotic devices."

Professor Sukkarieh is leading a three **phase program**, with the first **stage** using autonomous **perception** to enable robotic devices to read and understand their surroundings. He says the devices should be **commercially available** to farmers within the next couple of years.

With the support of **Horticulture** Australia, his team developed robotic systems, **sensors** and **intelligent** devices **trialled** on an **almond** farm in the regional center, Mildura.

"The robots can move through an **orchard** gathering data and developing a **comprehensive** in-ground and out-of-ground model of the entire orchard," says Professor Sukkarieh.

"Traditionally it has been necessary for someone to actually walk through the orchard, taking and analyzing soil and other samples and making decisions on the health and yield quality of the plants," he says.

"The devices we've developed can **collect**, analyze and **present** this information autonomously, so a major part of the farmer's job can be done automatically."

The second stage, which the team will **commence** in the new year, involves applying this technology to standard farm **tractors**, so that as well as being able to **perceive** their environment and identify any operations required, they will also be able to perform many of these operations themselves, such as applying fertilisers and pesticides, watering, **sweeping** and **mowing**.

The third and most **complex** stage will be to enable the devices to carry out harvesting.

"The devices we've developed already can identify each individual fruit on the tree and its degree of **ripeness**, which is about 80 percent of the job done. But being able to harvest them is our **ultimate** goal." As well as developing the technology, the team is working with farmers to **determine** how small changes to traditional agricultural practices can allow them to make the most of this new technology.

Vocabulary

automation [ˌɔːtəˈmeɪʃn] *n.* 自动化
phase [feɪz] *n.* 阶段
program [ˈprəʊgræm] *n.* 程序
stage [steɪdʒ] *n.* 阶段
perception [pəˈsepʃn] *n.* 知觉；收获
commercially [kəˈmɜːʃəli] *adv.* 商业上
available [əˈveɪləbl] *adj.* 可获得的；有空的
horticulture [ˈhɔːtɪkʌltʃə(r)] *n.* 园艺,园艺学
sensor [ˈsensə(r)] *n.* 传感器
trial [ˈtraɪəl] *n.* 试验 *v.* 试验；试用；测试(能力、质量、性能等)
almond [ˈɑːmənd] *n.* 杏树
orchard [ˈɔːtʃəd] *n.* 果园
comprehensive [ˌkɒmprɪˈhensɪv] *adj.* 广泛的；综合的
collect [kəˈlekt] *v.* 收集；收藏
present [prɪˈzent] *v.* 展现,显示
commence [kəˈmens] *v.* 开始；着手
tractor [ˈtræktə(r)] *n.* 拖拉机
perceive [pəˈsiːv] *v.* 意识到；察觉
sweep [swiːp] *v.* 打扫,清理；扫除
mow [məʊ] *v.* 割
complex [ˈkɒmpleks] *adj.* 复杂的；难懂的
ripeness [raɪpnəs] *n.* 成熟
ultimate [ˈʌltɪmət] *adj.* 最后的
determine [dɪˈtɜːmɪn] *v.* 决定

Notes

1. automated agriculture：农业自动化,是指应用自动控制和电子计算机等技术手段实现农业生产和管理的自动化,是农业现代化的标志之一。20 世纪 70 年代以来,农业逐渐推广应用自动控制、电子计算机、系统工程、遥感等技术,实现部分生产作业和管理自动化,取得了提高作物产量、效率和安全、省力等效果。农业自动化主要包括耕种、栽培、收割、运输、

排灌、作物管理、畜禽饲养等过程和温室的自动控制和最优管理。

2. “We can use automation to increase efficiency and yield, by having many of the manual tasks of farming performed by specially designed agricultural robotic devices.”

“我们可以用自动化来提高效率和增加产量,通过许多专门为农业设计的农业机械设备在农场完成手动操作才能完成的任务。”

field:产量; the manual tasks:手工活

3. With the support of Horticulture Australia, his team developed robotic systems, sensors and intelligent devices trialled on an almond farm in the regional center, Mildura.

在澳大利亚园艺学家的支持下,他的团队在米尔迪拉中心区域的杏仁农场进行试验开发了机器人系统、传感器和智能设备。

4. The robots can move through an orchard gathering data and developing a comprehensive in-ground and out-of-ground model of the entire orchard.

机器人可以穿过果园采集数据,并能开发出综合的包含地面地下的果园整体模型。

move:移动;gathering data and developing…为非谓语结构

5. The devices we've developed can collect, analyze and present this information autonomously, so a major part of the farmer's job can be done automatically.

我们开发的设备可以自主收集、分析和呈现这些信息,所以农民的主要工作都可以自动完成。

we're developed 前省略了引导词 that,做名词 the devices 的定语,是定语从句。

6. The second stage, which the team will commence in the new year, involves applying this technology to standard farm tractors, so that as well as being able to perceive their environment and identify any operations required, they will also be able to perform many of these operations themselves, such as applying fertilisers and pesticides, watering, sweeping and mowing.

新的一年,该团队将开始项目的第二阶段,包括应用本技术标准化操作农用拖拉机,以及感知环境、识别任何需要的操作,它们也能够执行这些操作本身,如施肥和喷洒农药、浇水、清扫和除草。

主句主语为 the second stage, involves 是谓语动词,which 引导的部分是非限制性的定语从句。

7. As well as developing the technology, the team is working with farmers to determine how small changes to traditional agricultural practices can allow them to make the most of this new technology.

还有技术开发,团队正与农民合作来决定对传统农业进行怎样的小变化能允许他们最大限度地使用这项技术。

主句中 the team 是主语,is working 是谓语,to determine…是目的状语。

Exercises

Ⅰ. Fill in the blanks with the words given below. Change the forms where necessary.

harvest	decision	commence	lack	intelligent
efficiency	identify	ultimate	drive	available

1. With the Asia-Pacific region's __________ of arable land, water, and infrastructure countries in the region are looking toward Australia for farming and agriculture solutions.
2. "There is a big __________ at the moment to conceptualise the future of Australian agriculture in terms of a 'food bowl' supplying the vast Asian market," says Professor Sukkarieh.
3. This is where automation can help. We can use it to increase __________ and yield, by having many of the manual tasks of farming performed by specially designed agricultural robotic devices.
4. He says the devices should be commercially __________ to farmers within the next couple of years.
5. With the support of Horticulture Australia, his team developed robotic systems, sensors and __________ devices trialled on an almond farm in the regional center, Mildura.
6. "Traditionally it has been necessary for someone to actually walk through the orchard, taking and analysing soil and other samples and making __________ on the health and yield quality of the plants," he says.
7. The second stage, which the team will __________ in the new year, involves applying this technology to standard farm tractors, so that as well as being able to perceive their environment and identify any operations required, they will also be able to perform many of these operations themselves, such as applying fertilisers and pesticides, watering, sweeping and mowing.
8. The third and most complex stage will be to enable the devices to carry out __________.
9. The devices we've developed already can __________ each individual fruit on the tree and its degree of ripeness, which is about 80 percent of the job done.
10. But being able to harvest them is our __________ goal.

Ⅱ. Translate the following paragraphs into Chinese.

"The devices we've developed can collect, analyze and present this information autonomously, so a major part of the farmer's job can be done automatically."

The second stage, which the team will commence in the new year, involves applying this technology to standard farm tractors, so that as well as being able to perceive their environment and identify any operations required, they will also be able to perform many of these operations themselves, such as applying fertilisers and pesticides, watering, sweeping and mowing.

参考译文

自动化农业

"我们可以用自动化来提高效率和增加产量,通过许多专门为农业设计的农业机械设备在农场完成手动操作才能完成的任务。"

苏克瑞教授主持了一项由三个阶段组成的项目,第一阶段运用自主感应设备使机器人阅读和理解它们周围的环境。他说,未来几年内农民们就可以通过商业渠道获得该设备了。

在澳大利亚园艺学家的支持下,他的团队在米尔迪拉中心区域的杏仁农场进行试验开发了机器人系统、传感器和智能设备。

"机器人可以穿过果园采集数据,并能开发出综合的包含地面地下的果园整体模型,"苏克瑞教授说。

"传统上需要有人真实地穿过果园,采集和分析土壤及其他样品,并对植物的健康和产量、品质作出判断,"他说。

"我们开发的设备可以自主收集、分析和呈现这些信息,所以农民的主要工作都可以自动完成。"

新的一年,该团队将开始项目的第二阶段,包括应用本技术标准化操作农用拖拉机,以及感知环境、识别任何需要的操作,它们也能够亲自执行这些操作,如施肥和喷洒农药、浇水、清扫和除草。

第三个也是最复杂的阶段将会是使设备进行收获。

"我们研发的设备可以识别树上每个水果及其成熟度,这大约是所有工作的80%。但能够收获它们是我们的最终目标。"还有技术开发,团队正与农民合作来决定对传统农业进行怎样的小变化能允许他们最大限度地使用这项技术。

Unit 6
Modern Agriculture

> *Implement the rural revitalization strategy and build a beautiful and happy new countryside.*
>
> ——实施乡村振兴战略，建设幸福新农村。

Situational Dialogue:

Visiting a Vineyard

A: Good morning, Mr. Brown.

B: Good morning, Ms. Chen. I've heard a lot about your vineyard and it seems to have a good fame.

A: Thank you. The vineyard has a long history, but it has got its fastest development only in the last ten years.

B: I suppose you are growing the best species of grapes.

A: Yes, we are. They're chosen to suit our climate conditions.

B: Most grapes are used in making wine, aren't they?

A: That's true. Only a small part is sold fresh and dried as raisins.

B: Is it hard to grow grapes?

A: Not really. But they require considerable care. The tendrils should be supported as they grow, and vines should be pruned regularly.

B: Do you create new varieties by grafting?

A: Sometimes, but in modern viticulture we often grow pollinated grapes to produce new varieties.

B: Does the farm bring you big profits?

A: Yes, it does.

raisin：葡萄干
tendril：藤蔓，葡萄藤
prune：修剪
graft：嫁接
viticulture：葡萄栽培

Lesson 1 Agriculture Mechanization

Engineering began to affect the farmer late in the 19th century, with steam-powered tractors and various tools for **drilling** seed holes and planting. Still, most fieldwork was done with hand tools like the **spade**, **hoe**, and **scythe**, or with hand-or animal-driven plows. A farmer's day was labor-intensive, beginning well before sunrise and ending at sunset.

Mechanization did not advance rapidly until the 20th century, with the **advent** of the internal combustion engine. As the chief power source for **vehicles**, it began replacing both horses and steam for planting, **cultivating**, and harvesting equipment. It made the **evolution** of the tractor possible, and led to sweeping changes in agriculture.

Major changes in tractor design include the power takeoff, the all-purpose or **tricycle**-type tractor, which enabled farmers to cultivate planted crops mechanically; rubber **tires**, which **facilitated** faster operating speeds; **treads** that could **negotiate** soft soil without getting stuck; and the switch to four-wheel drive and **diesel** power in the 1950s and 1960s, which greatly increased the tractor's pulling power. More recent innovations have led to the development of enormous tractors that can pull several **gangs** of plows while electronic systems **monitor** or control almost all of the power functions.

A large number of fatal **injuries** from tractors tipping over led to the design of **rollover** bars. They became commercially **available** in 1956 and later evolved into cabs, which provide a protective zone for operators, noise control, and a comfortable environment.

Engineering design for planting and harvesting was **hampered** by the wide variety of crops, all with different shapes and **consistencies** (e. g., corn, soybeans, wheat, cotton, and tomatoes). Nevertheless, an amazing array of innovations **peppered** the century, such as tractor-attachable cultivators and harvesters. Self-tying hay and straw balers arrived in 1940 along with a spindle cotton picker. **Shielded** corn-snapping rolls were developed in 1952, and **rotary** and **tine** separator combines were introduced in 1976, each reducing labor significantly.

A major **necessity** on many farms is a way to control soil erosion and reduce the time and energy to prepare **seedbeds**. The development of **chisel** and **disc tillage** tools and no-till planters in the 1970s and 1980s solved these problems. Even in the 1940s, sweep plows **undercut** wheat **stubble** to reduce wind and water erosion and **conserve** water.

Vocabulary

mechanization [ˌmekənaɪˈzeɪʃn] *n.* 机械化
engineering [ˌendʒɪˈnɪərɪŋ] *n.* 工程(学)
drill [drɪl] *v.* 钻(孔); 打(眼)
spade [speɪd] *n.* 铁锹,铲子
hoe [həʊ] *n.* 锄头
scythe [saɪð] *n.* (长柄)大镰刀
advent [ˈædvent] *n.* 出现; 到来
vehicle [ˈviːəkl] *n.* 车辆; 交通工具
cultivate [ˈkʌltɪveɪt] *v.* 耕作,种植
evolution [ˌiːvəˈluːʃn] *n.* 演变; 进化; 发展
tricycle [ˈtraɪsɪkl] *n.* 三轮车
tire [ˈtaɪə(r)] *n.* 轮胎
facilitate [fəˈsɪlɪteɪt] *v.* 促进,助长
tread [tred] *n.* 踏板
negotiate [nɪˈgəʊʃieɪt] *v.* 通过,越过
diesel [ˈdiːzl] *n.* 柴油机
gang [gæŋ] *n.* (工具,机械等的)一套
monitor [ˈmɒnɪtə(r)] *v.* 监控
injury [ˈɪndʒəri] *n.* 伤害
rollover [ˈrəʊləʊvə(r)] *n.* 翻车
available [əˈveɪləbl] *adj.* 有资格的
hamper [ˈhæmpə(r)] *v.* 妨碍,束缚
consistency [kənˈsɪstənsi] *n.* 连贯; 符合
pepper [ˈpepə(r)] *v.* 使布满

shielded [ˈʃiːldɪd] *adj.* 防护的
rotary [ˈrəʊtəri] *adj.* 旋转的
tine [taɪn] *n.* 尖头；齿；叉
necessity [nəˈsesəti] *n.* 必要，必需的事物
seedbed [ˈsiːdbed] *n.* 苗床
chisel [ˈtʃɪzl] *n.* 凿子
disc [dɪsk] *n.* 圆盘
tillage [ˈtɪlɪdʒ] *n.* 耕耘，耕地
undercut [ˌʌndəˈkʌt] *v.* 根除
stubble [ˈstʌbl] *n.* 残茎，茬子
conserve [kənˈsɜːv] *v.* 保护，保藏

Notes

1. internal combustion engine 内燃机；发动机
2. power source 电源，能源
3. sweeping change 彻底变化，全面改革
4. power takeoff 动力输出
5. all-purpose 通用的，多用途的
6. four-wheel drive 四轮驱动
7. a large number of 很多
8. tip over 使翻倒
9. straw baler 秸秆压捆机
10. cotton picker 采棉机
11. soil erosion 土壤侵蚀

Exercises

Ⅰ. Match the items listed in the following columns.

1. spade	A. 拖拉机
2. scythe	B. 三轮车
3. tricycle	C. 麦茬
4. wheat stubble	D. 镰刀
5. soil erosion	E. 铁锹
6. hoe	F. 锄头
7. tractor	G. 土壤侵蚀

Ⅱ. Translate the following sentences into English.

1. 农业机械化大大地提高了农场效率和生产力。

2. 农民日出而作、日落而息，劳动强度很大。

3. 大量因拖拉机翻倒而致死的伤亡事故促进了防翻滚障碍的设计。

4. 由于作物的品种繁多，玉米、大豆、小麦和马铃薯等作物的外形和坚硬程度各异，为作物的种植和收获而设计的机器设备常常受到限制。

参考译文

农业机械化

19 世纪晚期，工程学以创造了蒸汽动力的拖拉机以及其他各种钻孔和种植工具开始影响农业。然而，大部分的农活还是由铁锹、锄头和镰刀等手工工具或由人力拉犁或畜力拉犁来完成的。农民日出而作、日落而息，劳动强度很大。

直到 20 世纪，随着内燃机的出现，机械化才快速发展起来。内燃机作为拖拉机的主要动力来源，开始取代马匹和蒸汽动力用于种植、耕作和收获农作物。内燃机使得拖拉机的改良成为可能，引起了农业生产的巨大变革。

拖拉机设计发生了巨大的变化，如动力输出轴、多功能拖拉机或三轮拖拉机等设计使得农场主能机械化耕作所种植的作物；橡皮轮胎有助于加快拖拉机的运转速度，胎面可以使拖拉机通过松软的地面而不至于下陷；20 世纪 50 年代和 60 年代，四轮驱动及柴油动力大大提高了拖拉机的牵引力。近期以来的创新促进了大型拖拉机的发展，大型拖拉机通过电子系统操控，可以拉动好几组犁。

大量因拖拉机翻倒而致的伤亡事故促进了防翻滚障碍的设计。这种设计于 1956 年开始投入市场，后来发展成能为操作者提供防护空间、噪声控制和舒适环境的驾驶室。

由于作物的品种繁多，玉米、大豆、小麦、棉花和马铃薯等作物的外形和坚硬程度各异，为作物的种植和收获而设计的机器设备常常受到限制。尽管如此，20 世纪仍涌现出大批令人惊异的发明，如可连接拖拉机的耕田机和收割机。1940 年，自动打捆的干草和秸秆压捆机及纺锤形采棉机同时问世。1952 年问世的防护型玉米采收器和 1976 年问世的旋转齿圈脱粒联合收割机大大降低了劳动强度。

控制土壤侵蚀和减少准备作物苗床的时间和精力是许多农场必不可少的举措。20 世纪 70 年代和 80 年代出现的凿子、圆盘耕作工具和免耕式播种机等解决了这些问题。而在 20 世纪 40 年代，曲犁通过根切麦茬来减少风力和水力侵蚀并蓄养水源。

Lesson 2 Organic Farming

Organic farming is work **intensive**, in comparison with **mechanical** agriculture. It aims at the key of **high-yield**, high-returning, safety, sustainable farming of the future. Organic farming is one of the fastest growing **segments**.

Organic farming is best known for using natural **alternatives** to pesticides, such as natural **predators** and **rotating crops**. Farm owners are now discovering new methods to increasing food, like **diversifying** plants, using organic manure, and locating choices to pesticides. Organic food is more labor intensive since the farmers do not use pesticides, chemical fertilizers or drugs. Organic farming is the practice of producing food without the use of man-made pesticides, herbicides, and fertilizers. Organic farming is an environmentally responsible **approach** to producing high-quality food and **fiber**.

Organic farming can be the answer to the food security problems. Organic farming is a method used in farming without the use of any chemicals or **synthetics**. It is the **strategy** by which agriculturists create and develop vegetables, dairy foods, grains, meats and fruits. Organic farming is not only **beneficial** for farmers, but it also has proved useful for the dairy industry.

Organic farming was established in a bottom-up process as farmers aimed to design sustainable ways of using natural resources. It allows you support the nutrient sources from the soil and protect against land **smog** through chemical type **toxic contamination**. Organic farming helps in building richer soil. Organic farming helps keep our water supplies clean by stopping that **polluted runoff**. It can make the identical yields of corn or soybeans as traditional farming, for a 30% cheaper.

It can raise income for a time by reducing some time to make certain plants. It is a environmentally friendly farming method who makes healthful **vegetation** and animals with no damage to the planet. It is a **guaranteeing** farm **technique** with great results on the human ecological and public **atmosphere**.

Organic farming is a type of farming relying on such techniques like plant rotation, natural plant foods and biological pest control. Organic farming is more sustainable because it uses crop rotation to keep the soil rich. Organic farming has strong environmental benefits for soil and water quality, climate change **mitigation**, and **biodiversity**. Organic farming is a type of agriculture that benefits from the recycling and use of natural products. Organic farming has proved to be more cost-effective and eco-friendly than conventional farming.

Vocabulary

organic [ɔːˈgænɪk] *adj.* 有机(体)的
intensive [ɪnˈtensɪv] *adj.* 强烈的
mechanical [məˈkænɪkl] *adj.* 机械的
high-yield [ˈhaɪˈjiːld] *adj.* 产量很高的
segment [segˈmənt] *n.* 部分
alternative [ɔːlˈtɜːnətɪv] *n.* 可供选择的事物
predator [ˈpredətə(r)] *n.* 食肉动物
rotating crops 轮耕的作物
diversify [daɪˈvɜːsɪfaɪ] *v.* 使多样化,多样化
approach [əˈprəʊtʃ] *n.* 方法
fiber [ˈfaɪbə] *n.* 纤维
synthetic [sɪnˈθetɪk] *n.* 合成物
strategy [ˈstrætədʒi] *n.* 策略,战略
beneficial [ˌbenɪˈfɪʃl] *adj.* 有利的,有益的
smog [smɒg] *n.* 烟雾;烟尘
toxic contamination 有毒污染物
pollute [pəˈluːt] *v.* 污染
runoff [ˈrʌnˌɔːf] *n.* 径流
vegetation [ˌvedʒəˈteɪʃn] *n.* 植物(总称),草木

guarantee [ˌgærən'tiː] *v.* 保证,担保
technique [tek'niːk] *n.* 技巧; 技能
atmosphere ['ætməsfɪə(r)] *n.* 大气
mitigation [ˌmɪtɪ'geɪʃn] *n.* 缓解,减轻
biodiversity [ˌbaɪəʊdaɪ'vɜːsəti] *n.* 生物多样性

Notes

1. It aims at the key of high-yield, high-returning, safety, sustainable farming of the future.
   它以高产、高回报、安全可靠和对未来的可持续为目标。
2. Farm owners are now discovering new methods to increasing food, like diversifying plants, using organic manure, and locating choices to pesticides.
   农场主现在正在寻找新的方法来增加产量,像种植多样化作物、使用有机肥料和定位选择杀虫等。
   organic manure:有机肥料;pesticide:杀虫剂
3. It is the strategy by which agriculturists create and develop vegetables, dairy foods, grains, meats and fruits.
   它是由农民创造和发展出的生产蔬菜、奶制品、肉类和水果的策略。
4. Organic farming is not only beneficial for farmers, but it also has proved useful for the dairy industry.
   有机农业不仅有利于农民,事实证明它对乳制品行业也有利。
   not only… but also…:不但……而且; the dairy industry:乳品业
5. Organic farming was established in a bottom-up process as farmers aimed to design sustainable ways of using natural resources.
   由于有机农业以建立农民对自然资源的可持续利用为目标,所以是一个自下而上的过程。
   buttom-up:自下而上的; aim to:以……为目标
6. It is a guaranteeing farm technique with great results on the human ecological and public atmosphere.
   这是一种对人类生态和公共环境都有好处的保障性农业。
7. Organic farming is a type of farming relying on such techniques like plant rotation, natural plant foods and biological pest control.
   有机农业是依靠诸如农作物轮作、纯天然种植粮食和运用生物手段控制虫害等技术的农业。
   rely on:依靠,依赖; plant rotation:作物轮作

Exercises

Ⅰ. Put the following terms into English.

1. 有机农业
2. 生物多样性
3. 杀虫剂
4. 有机肥
5. 创新
6. 轮作
7. 可持续的
8. 乳品业
9. 蔬菜
10. 自然资源

Ⅱ. Put the following terms into Chinese.

1. corn
2. soybean
3. smog
4. scythe
5. advent
6. internal combustion engine
7. straw baler
8. grain
9. dramatically
10. yield

Ⅲ. Fill in the blanks with the words or expressions given below. Change the forms where necessary.

pollute	strategy	intensive	benefit	comparison with
sustainable	approach	prove	alternative	income

1. Organic farming is work ______________, in comparison with mechanical agriculture.
2. Your conclusion is wrong in ________________ their conclusion.
3. Organic farming is best known for using natural ____________ to pesticides, such as natural predators and rotating crops.
4. Homework—lots of it—is at the heart of her ____________ to forming opinions and to arguing them.
5. It is the ____________ by which agriculturists create and develop vegetables, dairy foods, grains, meats and fruits.
6. Organic farming is not only beneficial for farmers, but it also has ____________ useful for the dairy industry.
7. Organic farming helps keep our water supplies clean by stopping that ________ runoff.
8. It can raise ____________ for a time by reducing some time to make certain plants.
9. Organic farming is more ____________ because it uses crop rotation to keep the soil rich.
10. Organic farming has strong environmental ________ for soil and water quality, climate change mitigation, and biodiversity.

Ⅳ. Read the text and answer the following questions.

1. What is organic farming?

2. What is the aim of the organic farming?

3. Organic food is more labor intensive. How to understand this?

4. Which techniques is organic farming relying on?

Ⅴ. Supplemental reading.

Try to explain what is green manure and how to conserve soil in your own words.

What Is Green Manure?

Green manure is a valuable, natural fertilizer and soil conditioner（改良剂）and has an important part to play in organic gardening. It is a balanced mixture of silage（储藏的饲料）grains（颗粒）which when sprinkled（洒）on your soil from spring to late summer very quickly grows into a lush（茂密的）green mass（大块，大片）of young plants（takes only 36 weeks depending on the season）. These plants when dug（掘）in will rot（腐烂）down quickly to add valuable nutrients and vegetable matter to your soil, both feeding（增加营养）it and improving its structure.

Soil Conservation

Soil conservation（水土保持）protects soil from wind and water that can blow or wash it away. Good soil produces food crops for both people and animals.

Comprehensive（全面的）soil conservation is more than just the control of erosion（侵蚀）. It also includes the maintenance（维护，留住）of organic matter and nutrients in soil. Soil conservation practices also prevent the buildup（积累）of toxic（有毒的）substances（物质）in the soil, such as salts and excessive amounts of pesticides. Soil conservation maintains or improves soil fertility（肥力）, as well as its tilth（耕作层）, or structure. These all increase the capacity（能力）of the land to support the growth of plants on a sustainable basis.

There are two basic approaches to soil erosion control: barrier（阻挡）and cover（覆

盖). The barrier approach uses banks (堤) or walls such as earthen (泥制的) structures, grass strips (带), or hedgerows (灌木篱墙) to check (阻止) runoff (径流), wind velocity (速度), and soil movement.

Barrier techniques are commonly used all over the world. The cover approach maintains a soil cover of living and dead plant material. This cover lessens (减弱) the impact and runoff of rain water, and decreases the amount of soil carried with it.

This may be done through the use of cover crops, mulch (盖土), minimum tillage (耕作), or agroforestry (复合农林业).

参考译文

有机农业

与农业机械化相比较,有机农业是劳动密集型的。它以高产、高回报、安全可靠和对未来的可持续为目标。有机农业是农业中增长最快的一个领域。

有机农业因使用天然种植方法,如天然捕食者和作物轮作来替代农药而为人们所熟知。农场主现在正在寻找新的方法来增加产量,像种植多样化作物、使用有机肥料和定位选择杀虫等。由于有机农业中农民不使用农药、化肥或药物,其劳动力更加密集。有机农业不使用人造的农药、除草剂、肥料的食物生产。有机农业是对环境负责的生产高品质的食品和纤维的生产方式。

有机农业是解决食品安全问题的答案。有机农业是在农业中不使用任何化学品或合成材料的方法。它是由农民创造和发展出的生产蔬菜、奶制品、谷类、肉类和水果的方法。有机农业不仅有利于农民,事实证明它对乳制品行业也有利。

由于有机农业以建立农民对自然资源的可持续利用为目标,所以是一个自下而上的过程。它支持来自土壤的营养,防止有毒化学污染物进入土壤。有机耕作有助于建立更肥沃的土壤。有机耕作通过阻止被污染的径流而使水源保持清洁。它可以减少30%的用水而产出与传统农业相同产量的玉米或大豆。

它通过减少某种作物的种植时间在某段时间内增加收入。这是一种环境友好型的耕作方法,能够种植和饲养出既对地球无任何破坏又使人健康的作物和动物。这是一种对人类生态和公共环境都有好处的保障性农业。

有机农业是依靠诸如农作物轮作、纯天然种植粮食和运用生物手段控制虫害等技术的农业。因为采用轮作来保持土壤肥力,因此有机农业更具可持续性。有机耕种有益于土壤和水质的环境效益、减缓气候变化、生物多样性强。有机农业是从一种循环和运用自然产品中受益的农业。事实已经证明有机农业比传统的农业更具成本效益和环保效益。

Lesson 3 Agricultural Supply and Demand

Although modern agriculture is much more **efficient** and more food can be produced today than ever before, sometimes there are **drawbacks**. Many farmers use **chemicals** to **improve** their **produce**. They use **fertilizers** to help **crops** grow and **herbicides** and **pesticides** to kill **weeds** and **insects** that can **damage** crops.

Another problem with modern farming is that sometimes farmers grow so much grain that they cannot sell it at a profit. Sometimes governments buy it and store it. Governments try to keep the **supply** and **demand balanced** so that farmers do not grow too much and lose money or do not grow enough and make food prices rise **out of control**. It is a **challenge** to balance food supplies to **accommodate surpluses** (areas with too much food) and **shortages** (areas with too little food).

Vocabulary

efficient [ɪˈfɪʃnt] *adj.* 有效率的

drawback [ˈdrɔːbæk] *n.* 缺点

chemicals [ˈkemɪkl] *n.* 化学药品

improve [ɪmˈpruːv] *v.* 提高；改进

produce [prə'djuːs]　*n.* 产品；产量
fertilizer ['fɜːtəlaɪzə(r)]　*n.* 肥料，化肥
crop [krɒp]　*n.* 农作物；庄稼
herbicide ['hɜːbɪsaɪd]　*n.* 除草剂
pesticide [pes'tisaɪd]　*n.* 杀虫剂
weed [wiːd]　*n.* 杂草
insect ['ɪnsekt]　*n.* 虫，昆虫
damage ['dæmɪdʒ]　*v.* 损害
supply [sə'plaɪ]　*n.* 供给物　*v.* 供给；补充
demand [dɪ'mɑːnd]　*n.* 需求　*v.* 要求，请求；需要
balance ['bæləns]　*v.* （使）平衡　*n.* 平衡；天平
out of control　失控；不受控制
challenge ['tʃæləndʒ]　*n.* 挑战；盘问　*v.* 质疑；提出挑战
accommodate [ə'kɒmədeɪt]　*v.* 容纳；使适应
surplus ['sɜːpləs]　*n.* 剩余额
shortage ['ʃɔːtɪdʒi]　*n.* 不足

Notes

1. Although modern agriculture is much more efficient and more food can be produced today than ever before, sometimes there are drawbacks.
   尽管现代农业更高效，今天生产的食物比以往任何时候都多，有的时候现代农业也有缺点。
2. They use fertilizers to help crops grow and herbicides and pesticides to kill weeds and insects that can damage crops.
   他们使用化肥来帮助农作物生长，使用可能会破坏农作物的除草剂和杀虫剂来杀死杂草和昆虫。
3. Another problem with modern farming is that sometimes farmers grow so much grain that they cannot sell it at a profit.
   现代农业的另一个问题是，有时农民种植的粮食如此之多，以至于他们不能在盈利的基础上销售。
4. Governments try to keep the supply and demand balanced so that farmers do not grow too much and lose money or do not grow enough and make food prices rise out of control.
   政府试图保持供求平衡，以使农民不至于种得过多而赔钱或种得过少致使食品价格失控上涨。

Exercises

Ⅰ. Fill in the blanks with the words or expressions given below. Change the forms where necessary.

balance	efficient	reason	conservation	common
out of control	damage	improve	profit	instead of

1. Although modern agriculture is much more ________ and more food can be produced today than ever before, sometimes there are drawbacks.
2. Many farmers use chemicals to ________ their produce.
3. They use fertilizers to help crops grow and herbicides and pesticides to kill weeds and insects that can ________ crops.
4. They use animal manure ________ chemical fertilizers and herbal remedies instead of antibiotics.
5. Another problem with modern farming is that sometimes farmers grow so much grain that they cannot sell it at a ________.
6. Governments try to keep the supply and demand balanced so that farmers do not grow too much and lose money or do not grow enough and make food prices rise ________.
7. It is a challenge to ________ food supplies to accommodate surpluses (areas with too much food) and shortages (areas with too little food).
8. Famine, starvation, and malnutrition are ________ problems in these countries.
9. Farmers in these countries need to produce their own food and adopt land ________ practices that allow them to farm more productively.
10. Sometimes the ________ for shortages are financial, but in other instances, farmers just lack the proper education.

Ⅱ. Translate the following paragraph into Cninese.

In less-developed, poorer countries, there is a different problem: not enough food to feed the population. Famine, starvation, and malnutrition are common problems in these countries. Sometimes, other countries will send emergency food supplies to these nations, but this is not a long-term solution. Farmers in these countries need to produce their own food and adopt land conservation practices that allow them to farm more productively. Sometimes the reasons for shortages are financial, but in other instances, farmers just lack the proper education.

Ⅲ. Supplemental reading.

Read the following passage and explain what is organic farming and crop rotation in your own words. Don't forget to combine the knowledge that you've learned with the main idea of the unit.

Organic Farming & Crop Rotation

Organic farming can be defined as an approach to agriculture where the aim is to create integrated (整体的), humane (人性化的), environmentally and economically sustainable agricultural production systems. Organic farming is a form of agriculture that relies on crop rotation, green manure, compost (堆肥), biological pest control, and mechanical cultivation to maintain soil productivity and control pests.

Crop rotation is to grow specific groups of vegetables on a different piece of land each year. Groups are moved around in sequence (序列), so they don't return to the same spot for at least three years. Crops are rotated to prevent exhausting soil nutrients while adding certain matters necessary for the following crops. Some plants have so few soil dwelling (生长在土壤里的) pests or disease that they serve as effective pest management tools in the rotation.

Pest control aims to control animal pests (including insects), fungi, weeds and disease. Organic pest control involves many techniques (技巧), including, allowing for an acceptable level of pest damage, encouraging or even introducing beneficial organisms, careful crop selection and crop rotation, and mechanical controls such as traps (陷阱).

For weed elimination (根除), the traditional method is to remove weeds by hand, which is too costly in developed countries where labor is more expensive. One recent innovation (创新) in rice farming is to introduce ducks and fish to wet paddy fields (稻田), which eat both weeds and insects. Another effective and organic way of eliminating weeds is mechanical cultivation, but it is labor consuming. Pestresistant and genetically modified crops have been proposed as an alternative (替代物) to pesticide use.

The organic movement began in the 1930s and 1940s as a reaction to agriculture's growing reliance on synthetic fertilizers. Artificial fertilizers had been created during the 18th century.

These early fertilizers were cheap, powerful, and easy to transport in bulk (成批). The 1940s is referred to as the "pesticide era".

Organic farming has remained tiny (微小的) since its beginning, and organic output remains small, but it has been growing rapidly in many countries, notably (尤其, 显著) in Europe.

参考译文

农业供需

尽管现代农业更高效,今天生产的食物比以往任何时候都多,有的时候现代农业也有缺点。许多农民使用化学药品来增加生产。他们使用化肥来帮助农作物生长,使用可能会伤害农作物的除草剂和杀虫剂来杀死杂草和害虫。

现代农业的另一个问题是,有时农民种植的粮食如此之多,以至于他们不能在盈利的基础上销售。有时政府购买并储存这些剩余的粮食。政府试图保持供求平衡,以使农民不至于种得过多而赔钱或种得过少致使食品价格失控上涨。平衡粮食供应使剩余(粮食过多的地区)和缺乏(粮食生产不足的区域)之间保持平衡是相当有挑战的。

Unit 7 Planting

> *Plant a grain of millet in spring and harvest ten thousand seeds in autumn.*
> ——春种一粒粟，秋收万颗子。

Situational Dialogue:

Export of Frozen Shrimps

A: Yesterday we discussed with you about the possibility of exporting your products to United States. Now we'd like to hear what you would say.

B: We greatly appreciate your interest in our products.

A: Then we would like to have an idea of your pricing strucure for the frozen shrimps.

B: The prices are of course very much dependent on the quantity pruchased by our customers.

A: We understand that.

B: The basic export price for our frozen shrimps is \$4.75 per kg. We normally ship on an FOB basis. Small quantities are sometimes shipped by air freight; however, large quantities are shipped by ocean freight.

A: What are your major export markets?

B: Japan is our major purchaser, followed by the EU. Japan imports 80% of our various frozen products.

A: We are also very much interested in purchasing your frozen shrimps and we'd like to visit your company to see how they are processed.

B: No problem. I'll ask my staff to make arrangements for a visit to the company this afternoon. Shall we now break for lunch?

A: It sounds like an excellent idea.

frozen shrimp：冰虾
export：出口
FOB：离岸价格(Free on Board)

Lesson 1 Onion Planting

It is hard to say when the onion came into being. They were grown in ancient Egypt, and **eventually** arrived in Rome and became known as the word onion.

Selecting the Best Varieties for Your Area

When onions are first planted, their growth is **concentrated** on new roots and green leaves or tops. The onion will first form a top and then when a **specific combination** of daylight, darkness, and **temperature** is reached, **bulb** formation starts. The size of the **mature** onion bulb is dependent on the number and size of the tops. For each leaf, there will be a ring of onion; the larger the leaf, the larger the ring will be. For instance, a variety that needs many hours of summer light will not **perform** well in an area that receives fewer hours of light. Onion growers **categorize** onions in one of three ways: Short Day, **Intermediate** Day, and Long Day.

Tips for Successful Onion Growth

Onion plants are **hardy** and can **withstand** temperatures as low as 20 ℉. They should be set out 4 to 6 weeks **prior** to the date of the last **average** spring **freeze**.

When you obtain onion plants, they should be dry. Do not wet them or **stick** their roots in soil or water. Unpack your plants and store them in a cool, dry place until you plant them. Properly stored onion plants will last up to three weeks. Do not worry if the plants become dry. As soon as they are planted, they will "shoot" new roots and green tops.

Before obtaining your plants, you may want to begin soil preparation. Onions are best grown on raised beds at least 4 **inches** high and 20 inches wide. Onions need a very fertile and well-balanced soil. Organic gardeners should work in rich finished **compost**, high in **nitrogen** and **phosphorus** with plentiful minerals. **Spread lime** if soil is too **acidic**. If using commercial fertilizer, make a **trench** in the top of the bed 4 inches deep, **distribute** one-half cup of the fertilizer per 10 feet of row. Cover the fertilizer with 2 inches of soil.

Cooking Tips

To reduce tearing when **peeling** or **slicing** an onion, **chill** for 30 minutes or cut off the top, but leave the root on. The root has the largest amount of **sulphuric compounds**, which is what causes tears when the onion is peeled or cut. Remove the root prior to cooking or eating.

Prolonged cooking takes the flavor out of onions. Cook only until they're tender when tested with a fork.

Nutritional Information

1 middle raw onion contains:

- 60 **calories**
- 1 gram **protein**
- 14 grams **carbohydrates**
- 0 **fat**
- 0 **cholesterol**
- 10 mg **sodium**
- 200 mg **potassium**
- 11.9 mg **vitamin C** (20% of USRDA)

Did You Know?

Onions are high in energy and water content. They are low in calories, and have a **generous** amount of B6, B1, and **folic acid**.

Onions contain chemicals which help fight the free **radicals** in our bodies. Free radicals cause disease and destruction to cells which are linked to at least 60 diseases.

To make onions milder, **soak** them in milk or pour boiling water over slices and let stand. **Rinse** with cold water.

When a person eats at least 1/2 a raw onion a day, their HDL cholesterol goes up an average of 30%. Onions increase **circulation**, lower blood pressure, and prevent blood clotting.

Vocabulary

eventually [ɪ'ventʃuəli] *adv.* 最后
variety [və'raɪəti] *n.* 多样;种类
concentrate ['kɒnsntreɪt] *v.* 浓缩;集中
specific [spə'sɪfɪk] *adj.* 具体的,特定的
combination [ˌkɒmbɪ'neɪʃn] *n.* 结合
temperature ['temprətʃə(r)] *n.* 温度
bulb [bʌlb] *n.* 球茎,块茎植
mature [mə'tʃʊə(r)] *adj.* 成熟的
perform [pə'fɔːm] *v.* 工作;做
categorize ['kætəgəraɪz] *v.* 把……归类
intermediate [ˌɪntə'miːdiət] *n.* 中间物,中间分子 *adj.* 中间的
hardy ['hɑːdi] *adj.* [植]耐寒的
withstand [wɪð'stænd] *v.* 经受,承受
prior ['praɪə(r)] *adj.* 优先的
average ['ævərɪdʒ] *n.* 平均水平
freeze [friːz] *n.* 冻结
stick [stɪk] *v.* 刺入;插入
inch [ɪntʃ] *n.* 英寸
compost ['kɒmpɒst] *n.* 堆肥
nitrogen ['naɪtrədʒən] *n.* [化]氮
phosphorus ['fɒsfərəs] *n.* [化]磷
spread [spred] *v.* 散布;播撒
lime [laɪm] *n.* 石灰
acidic [ə'sɪdɪk] *adj.* (味)酸的
trench [trentʃ] *n.* 沟,渠
distribute [dɪ'strɪbjuːt] *v.* 散布
peel [piːl] *v.* 剥皮
slice [slaɪs] *v.* 切片
chill [tʃɪl] *v.* 冷冻;冷藏
sulphuric [sʌl'fjʊərɪk] *adj.* 硫黄的
compound ['kɒmpaʊnd] *n.* 复合物
prolong [prə'lɒŋ] *v.* 延长
calory [kæləri] *n.* 卡路里(热量单位:calorie)
protein ['prəʊtiːn] *n.* 蛋白质

carbohydrate [ˌkɑːbəˈhaɪdreɪt] *n.* 碳水化合物
fat [fæt] *n.* 脂肪
cholesterol [kəˈlestərɒl] *n.* 胆固醇
sodium [ˈsəʊdiəm] *n.* [化]钠
potassium [pəˈtæsiəm] *n.* [化]钾
vitamin [ˈvɪtəmɪn] *n.* 维生素
generous [ˈdʒenərəs] *adj.* 丰富的,充足的
folic acid 叶酸
radical [ˈrædɪkl] *n.* 自由基
soak [səʊk] *v.* 浸泡
rinse [rɪns] *v.* 漂洗;冲洗
circulation [ˌsɜːkjəˈleɪʃn] *n.* 血液循环

Notes

1. It is hard to say when the onion came into being. They were grown in ancient Egypt, and eventually arrived in Rome and became known as the word onion.
   很难讲清洋葱是什么时候出现的。古埃及人种植洋葱,最终,洋葱传到罗马,并被称为"onion"。
   come into being:形成,产生
2. For each leaf, there will be a ring of onion; the larger the leaf, the larger the ring will be.
   每一片叶子会长出一个洋葱圈;叶子越大,洋葱圈就越大。
   a ring of onion 一个洋忽圈
3. Onions need a very fertile and well-balanced soil. Organic gardeners should work in rich finished compost, high in nitrogen and phosphorus with plentiful minerals. Spread lime if soil is too acidic.
   洋葱需要非常肥沃而且均衡的土壤。有机园艺师应该在富含充足氮磷等矿物质的成品堆肥上种植。如果土壤过酸的话就要撒上石灰。
4. Prolonged cooking takes the flavor out of onions. Cook only until they're tender when tested with a fork.
   延长烹调时间祛除洋葱的辣味。在烹调时用一把叉子试试,洋葱变软就熟了。
   take... out of:除去
5. Onions contain chemicals which help fight the free radicals in our bodies. Free radicals cause disease and destruction to cells which are linked to at least 60 diseases.
   洋葱含有可以帮助我们抵抗体内自由基的化学物质。这些自由基会引起疾病,破坏与至少 60 种疾病有关的细胞。
   the free radicals:自由基;destruction:破坏

Exercises

I. Read the text and answer the following questions.

1. How to select the best varieties for your area?

2. Can you give some tips for successful onion planting? Such as temperature, soil and onion plants (seeds)?

3. Do you know some tips for peeling and slicing onion?

4. Please give some suggestions for cooking onion.

5. Do you know some medical functions of onion? What are they?

II. Put the sentences in correct order for planting onion.

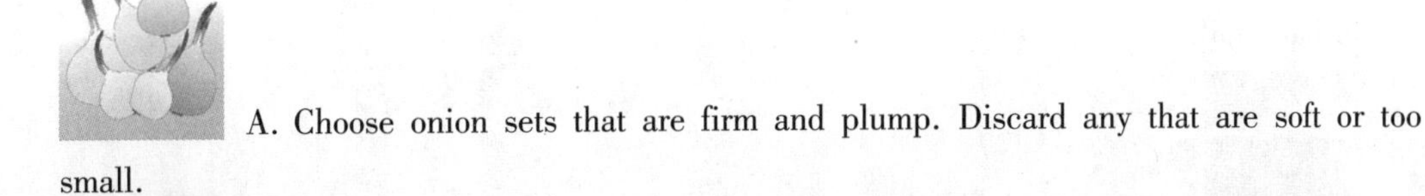

A. Choose onion sets that are firm and plump. Discard any that are soft or too small.

B. Use a trowel to dig small holes only as deep as each onion set so the tip is exposed when you replace the soil (about 1 inch). Gently firm down the soil around the tips with your fingers. The sets should be planted with the tips pointing upwards and positioned about 10 centimetres apart. Each row should be 20-30 centimetres apart.

C. Harvest the onions in late spring.

D. Loosen the soil with a fork and remove any weeds or large stones.

E. Water the onions in the spring, but you may not need to water them in autumn and winter.

F. Use a landscape rake to create a level surface. If you have poor soil you may want to add organic matter before you start sowing the sets.

G. Make rows of small holes in the soil. You can lay a piece of string on the soil as a guide to make sure you dig the row in a straight line.

H. Lightly rake the surface once more.

Ⅰ. Use your feet or the head of the rake to firm down the soil as onions grow well in hard soil.

J. Find a good location. Onions should be planted in a sunny or partially shady spot with little wind. They shouldn't be planted in heavy clay soil.

Ⅲ. Scan the following chart about soil types and then complete the given paragraph.

Soil Types	Components and Characteristics（成分和特性）
Sandy Soil（砂土）	This type has the biggest particles（微粒）. It consists of rock and mineral particles that are very small. Therefore the texture is gritty（含砂的）. Sandy soil is formed by the disintegration（分解）and weathering（侵蚀）of rocks such as limestone, granite, quartz and shale（石灰石、花岗石、石英和页岩）. The soil is easier to cultivate if it is rich in organic material but then it allows drainage（排水）more than what is needed, thus resulting in over-drainage and dehydration of the plants in summer. It warms very fast in the spring season.
Silty Soil（粉砂土）	It is considered to be one of the most fertile of soils. It consists of minerals like quartz（石英）and fine organic particles（微小的有机颗粒）. It has more nutrients than sandy soil and it also offers better drainage. In case silty soil is dry it has a smoother texture and looks like dark sand. It offers better drainage and is much easier to work with when it has moisture（湿气）.
Clay Soil（黏质土）	Clay is a kind of material that occurs naturally and consists of very fine grained（微小的颗粒）material with very less air spaces, that is the reason it is difficult to work with since the drainage in this soil is low. Clay soil becomes very heavy when wet and if cultivation has to be done, organic fertilizers have to be added.
Loamy Soil（壤质土）	This soil consists of sand, silt and clay to some extent. It is considered to be the perfect soil. The texture is gritty（含砂的）and retains water very easily, yet the drainage is well. There are various kinds of loamy soil ranging from fertile to very muddy and thick sod（草地）. Yet out of all the different kinds of soil loamy soil is the ideal for cultivation.
Peaty Soil（泥炭土）	This kind of soil is basically formed by the accumulation of dead and decayed organic matter, and it naturally contains much more organic matter than most of the soils. Now the decomposition of the organic matter in peaty soil is blocked by the acidity of the soil. Though the soil is rich in organic matter, nutrients present are fewer in this soil type than any other type. Peaty soil is prone to water logging（水浸）but if the soil is fertilized well and the drainage of the soil is looked after, it can be the ideal for growing plants.
Chalky Soil（白垩土）	Chalky soil is very alkaline in nature and consists of a large number of stones. The fertility of this kind of soil depends on the depth of the soil that is on the bed of chalk. This kind of soil is prone to dryness and in summer it is a poor choice for cultivation, as the plants would need much more watering and fertilizing than on any other type of soil.

The two soils that offer better drainage are 1. ________. The soil that offers worst drainage is 2. ________. The soil that is acidic and rich in organic matter is 3. ________. The soil that is very alkaline is 4. ________. And the ideal soil for cultivation is 5. ________.

参考译文

洋葱种植

很难讲清洋葱是什么时候出现的。古埃及人广泛种植洋葱,最终,洋葱传到罗马,并被称为“onion”。

选择你所在地域的最佳品种

洋葱种植之初,其生长主要集中在新长出的根和绿色的叶或顶端。洋葱会首先长出顶部,随后,当特定的日照、黑暗和温度达到了,洋葱球就开始形成。成熟的洋葱鳞茎的大小是依据顶尖的数量和大小。每一片叶子会长出一个洋葱圈;叶子越大,洋葱圈就越大。值得注意的是,每个品种都有特定的要求。例如,需要长时间光照的品种在光照较短的地方就不会长得很好。洋葱种植者把洋葱粉为三类:长时光照、短时光照和中度光照。

洋葱种植技巧

洋葱是耐寒植物,能经受温度低至 20 华氏度。在春季最后一次霜冻之前的 4 到 6 周就要播种。

要种植洋葱,洋葱应该干燥。不沾水或把洋葱的根插在土壤或水里。打开洋葱,把它们存储在阴凉、干燥的地方直到播种。妥善储存洋葱能保持三周。如果变得干燥也不要担心。一旦种植,它们将“抽出”新的根和绿色茎尖。

在种植洋葱前,你要先整好地。洋葱最好种植在 4 英寸高,20 英寸宽的苗床上。洋葱需要非常肥沃而且均衡的土壤。有机园艺师应该在富含充足氮磷等矿物质的成品堆肥上种植。如果土壤过酸的话就要撒上石灰。如果使用商业肥料的话,就在苗床上做一个 4 英寸深的沟槽,一行中每 10 英尺撒半杯肥料。肥料上盖 2 英寸的土。

烹饪技巧

在剥洋葱或切洋葱时,为了减少流泪,将其冻 30 分钟或者切断顶部,只留下洋葱根。洋葱根含有大量硫酸化合物,这就是剥或者切洋葱时流泪的原因。将洋葱根部以上的部位拿来烹调或者食用。

延长烹调时间祛除洋葱的辣味。在烹调时用一把叉子试试,洋葱变软就熟了。

营养信息

一个中等大小的生洋葱含有:

- 60 卡路里
- 1 克蛋白质
- 14 克碳水化合物
- 0 脂肪
- 0 胆固醇

- 10 毫克钠
- 200 毫克钾
- 11.9 毫克维生素 C(20% USRDA)

你知道吗?

洋葱的能量和水分含量很高。它们热量低,并富含大量的维生素 B6、B1 和叶酸。

洋葱含有可以帮助我们抵抗体内自由基的化学物质。这些自由基会引起疾病,破坏与至少 60 种疾病有关的细胞。

为使洋葱温和,将其浸泡在牛奶中或者把开水倒在洋葱片上,泡一会儿,然后用冷水冲洗。

当一个人一天吃至少半个生洋葱时,他们的高密度脂蛋白胆固醇平均上升 30%。洋葱增强血液循环,降低血压,防止血液凝固。

Lesson 2 Planting Guide for Garlic

Preparation

Garlic prefers sandy loam soil, but grows well in nearly any well-drained, slightly acidic (pH of 6~7), fertile soil. The looser the **composition** of the soil, the larger your garlic will grow. Prepare your garlic bed by turning under or **tilling** in compost (be sure to use compost that is fully **aerobically** broken down and contains **animal manures** and **plant residues**, rather than **cedar** or **redwood**). Make sure your soil has **ample** phosphorus. Avoid planting garlic in the same place you've previously grown garlic, onions or any other **alliums** for 3 years, because of the potential for spreading diseases like white rot. Gophers love garlic; protect your beds with **gopher** wire or traps.

Planting & Growing

You will receive whole garlic bulbs from us, but you will be planting the **cloves** (the sections of the bulb). First "crack" the garlic bulb (separate the cloves for planting). Once you have cracked the bulbs it is best to plant within 5~7 days before cloves begin to dry out. A clove pre-planting **dip** can help the cloves off to a good start: in one **gallon** of water, place two **tablespoon** of Liquid Kelp Extract & Humic

Acids. **Removal** of the clove skin cover is not necessary. Plant your garlic cloves root end down. Plant garlic cloves in rows spaced 18″ apart. Plant cloves 4″ to 6″ apart down the rows. Down Elephant garlic rows, plant 8″~9″ apart. Cover cloves with about 2″ of soil. After you have planted, it is a good idea to **mulch** the garlic. Leaves, compost or **straw** make good mulches. In the south, a light mulch will **suffice**, but in cold northern areas, up to 8″ of mulch is recommended. Remove the mulch in the spring once **frost** danger is over. Water well and then only water again when the soil is dry. Moisture is a **critical** factor in spring; watch your soil moisture levels and irrigate accordingly. Remember, garlic loves water and food, but it must have good **drainage** or it will rot. In the spring, feed the garlic with either composted manure or a well-balanced fertilizer before the bulbs begin to enlarge.

Keep the weeds away from your garlic at all times; you either grow weeds or garlic, but not both! If your garlic sends up a flower stalk in spring, **snip** it off, making the cut as close as feasible, or it might grow back. These flower stocks, called **scapes**, are delicious in **stir fry**. They have a mild garlic taste and the **texture** of green beans.

Harvesting & Storing

When the garlic leaves begin to turn yellow in the summer, stop irrigating for 2 weeks and then pull up the plant. Immediately place plants in a **shady** place to cure. Do not leave your garlic in the sun because it will sunburn and rot. A good way to store garlic is to tie it or **braid** it and hang it in a dark place where it will receive good air **circulation**.

Vocabulary

garlic [ˈɡɑːlɪk] *n.* 大蒜
composition [ˌkɒmpəˈzɪʃn] *n.* 成分；构成
till [tɪl] *v.* 耕作，犁地
aerobically [eəˈrəʊbɪkli] *adv.* 需氧地，有氧地
animal manure 动物粪便
plant residue 植物残体
cedar [ˈsiːdə(r)] *n.* 雪松
redwood [ˈredwʊd] *n.* 红杉
ample [ˈæmpl] *adj.* 足够的；充足的

allium [ˈəlɪəm] *n.* 葱属植物
gopher [ˈɡəʊfə(r)] *n.* 地鼠
clove [kləʊv] *n.* 丁香
dip [dɪp] *v.* 浸
gallon [ˈɡælən] *n.* 加仑
tablespoon [ˈteɪblspuːn] *n.* 大汤匙,大调羹
removal [rɪˈmuːvl] *n.* 免职;除去
mulch [mʌltʃ] *v.* 覆盖
straw [strɔː] *n.* 稻草;麦秆
suffice [səˈfaɪs] *vt.* 满足……的需要;使满足
frost [frɒst] *n.* 霜冻,结霜
critical [ˈkrɪtɪkl] *adj.* 关键的,极重要的
drainage [ˈdreɪnɪdʒ] *n.* 排水
snip [snɪp] *v.* 剪去
scape [skeip] *n.* 花茎
stir fry 用旺火煸
texture [ˈtekstʃə(r)] *n.* 质地;结构
shady [ˈʃeɪdi] *adj.* 背阴的
braid [breɪd] *v.* 把……编成辫子
circulation [ˌsɜːkjəˈleɪʃn] *n.* 流通;血液循环

Notes

1. Garlic prefers sandy loam soil, but grows well in nearly any well-drained, slightly acidic (pH of 6~7), fertile soil.
   大蒜喜欢沙质土壤,但在任何排水良好的微酸性(pH 值 6~7)肥沃的土壤里也能生长得很好。
   sandy loam soil 沙质土壤; well-drained 排水良好的; slightly acidic (pH of 6~7) 微酸性的(pH 值 6~7)
2. The looser the composition of the soil, the larger your garlic will grow.
   土壤越松散,大蒜长得越大。
3. Make sure your soil has ample phosphorus. Avoid planting garlic in the same place you've previously grown garlic, onions or any other alliums for 3 years, because of the potential for spreading diseases like white rot.
   确保你有足够的磷土壤。因为存在传染如白腐病等疾病的可能,所以 3 年里避免在以前种植大蒜的同一个地方种植大蒜、洋葱和任何其他葱蒜类蔬菜。

phosphorus 磷,磷光物质,含磷的

4. A clove pre-planting dip can help the cloves off to a good start: in one gallon of water, place two tablespoon of Liquid Kelp Extract & Humic Acids.
种植前将剥下的鳞茎沾水有助于它的生长:加一加仑的水,两汤匙的海带提取液或者腐殖酸。
5. Moisture is a critical factor in spring; watch your soil moisture levels and irrigate accordingly.
水分是春天里的一个关键因素;注意你的土壤水分,适时浇水。
6. Keep the weeds away from your garlic at all times; you either grow weeds or garlic, but not both! If your garlic sends up a flower stalk in spring, snip it off, making the cut as close as feasible, or it might grow back.
保持种植的大蒜始终没有杂草;要么种杂草,要么种大蒜,两者不能并存! 春天时,如果你种植的大蒜抽出了花枝,剪下来,剪得尽可能彻底,不然它会再长出来。

Exercises

Ⅰ. Put the following terms into Chinese.

1. sandy loam soil
2. well-drained
3. slightly acidic
4. fertile soil
5. compost
6. white rot
7. the relative humidity
8. straw
9. irrigate
10. weeds
11. clover
12. texture
13. dark place
14. good air circulation

Ⅱ. Transalte the sentences into Chinese.

1. Prepare your garlic bed by turning under or tilling in compost (be sure to use compost that is fully aerobically broken down and contains animal manures and plant residues, rather than cedar or redwood).

2. Plant garlic cloves (and Elephant garlic) in rows spaced 18″ apart. Plant cloves 4″ to 6″ apart down the rows.

3. Remember, garlic loves water and food, but it must have good drainage or it will rot. In the spring, feed the garlic with either composted manure or a well-balanced fertilizer before the bulbs begin to enlarge.

4. When the garlic leaves begin to turn yellow in the summer, stop irrigating for 2 weeks and then pull up the plant. Immediately place plants in a shady place to cure.

5. A good way to cure garlic is to tie it or braid it (if it is a soft neck garlic) and hang it in a dark place where it will receive good air circulation.

大蒜种植

准　备

大蒜喜欢沙质土壤,但在任何排水良好的微酸性(pH 值 6~7)肥沃的土壤里也能生长得很好。土壤越松散,大蒜长得越大。通过翻动堆肥或在堆肥上耕作来备好大蒜的苗床(一定要使用被彻底有氧分解,含有动物粪便及植物残体的堆肥,而不是杉木或红木)。确保你有足够的磷。因为存在传染如白腐病等疾病的可能,所以 3 年里避免在以前种植大蒜的同一个地方种植大蒜、洋葱和任何其他葱蒜类蔬菜。地鼠喜欢大蒜;防止老鼠洞毁坏大蒜苗床。

种植与生长

收到我们的蒜头(大蒜球茎),将小鳞茎(球茎部分)种下。首先"剥离"大蒜鳞茎(把要种植的鳞茎分离下来)。一旦把鳞茎剥离下来,最好在 5~7 天内就将它种下去,否则它就会变干。种植前将剥下的鳞茎沾水有助于它的生长:加一加仑的水,两汤匙的海藻萃取液及腐殖酸。去除不必要的表皮。把大蒜根部种下。以 18 英寸的间隔将大蒜种成行。表皮鳞茎每行间隔 4 到 6 英寸。独立的蒜头间隔 8 到 9 英寸。覆盖约 2 英寸厚的土壤。种植时给大蒜盖上地膜是一个好主意。叶、堆肥或稻草可做很好的覆盖物。在南方,薄的覆盖物就足够了，但在寒冷的北方地区，建议覆盖的厚度要达到 8 英寸。一旦春季霜冻的危险已经过去就揭去覆盖物。浇水，土壤干旱时再次浇水。水分是春天里的一个关键因素；注意你的土壤水分，适时浇水。记住，大蒜要多浇水和施肥，但必须有良好的排水，否则它会腐烂。春季在大蒜开始长大时给大蒜施腐熟的有机肥或营养均衡的化肥。

保持种植的大蒜始终没有杂草；要么种杂草，要么种大蒜，两者不能同时并存！春天时，如果你种植的大蒜抽出了花枝，剪下来，剪得尽可能彻底，不然它会再长出来。这些花枝，称为蒜苔，炒出来非常美味。它们既有温和的蒜味和又有绿豆的口感。

收获与储存

夏季当大蒜的叶子开始变黄的时候，两周不要浇水，然后把它拔下来。立刻将大蒜放置在阴凉处。不要把大蒜放在太阳底下，因为会晒焦或腐烂。正确的储存方法是将它编起来或者编织在一起（如果蒜瓣较软的话），然后把它悬挂在通风较好的阴凉处。

Lesson 3 Cucumber Production in Greenhouse

Planting

Cucumbers generally grow more rapidly than tomatoes and produce earlier. They also require higher temperatures, which means they are generally grown as a spring or early summer crop. Daytime temperatures should be 80-85 ℉ (nighttime 65-75 ℉). Soil temperatures should be at least 65 ℉. Lower temperatures will **delay** plant growth and fruit development.

Plants are best started in **individual** containers. As seed are often very expensive, **sow** one seed per **container** (1/4 to 1/2 inch deep). Water, cover pots with clear **polyethylene**, and place in the **shade**. Plants will **emerge** in two to three days at 80-85 ℉. Remove plastic coverings when plants emerge and place them in full sun. After plants have formed at least two true leaves, **transplant** them to their permanent location in the growing bed. Cucumbers will require 6-8 square feet of space per plant. Plants are generally spaced 2 feet **apart** in rows 3 to 4 feet apart.

Harvesting

With good **management**, each plant may produce as much as 20-30 pounds of fruit over a four-month period. European varieties are generally harvested when fruit are 12-16 inches long and 3/4-1 pound in size. Fruit are often **shrink-wrapped** to prevent softening from **moisture loss**. Store fruit at 55 ℉ with 80-90 percent relative humidity.

Seedless European greenhouse cucumbers are distinctly different from traditional field-grown cucumbers. Because of consumer **expectations** for field-grown cucumbers, greenhouse cucumbers may require some market promotion. Excellent selling points include their seedlessness, dark green color, **mild flavor**, and thin, tender skins that require no **peeling**.

Pest Control

Gummy stem blight is a **fungus** that occurs on all above-ground parts of the plant causing extensive damage to leaves, stems, and fruit. Light brown to black **lesions** occur on leaves, at **nodes**, and in **pruning wounds**. Leaf lesions eventually dry and fall from leaves. Stem lesions can **crack** at the soil line, producing an amber-colored **gummy ooze**, and can **girdle** the plant resulting in death. This **disease** also can occur as **grayish-green water-soaked** lesions on fruit beginning at the **blossom** end and can develop on fruit after harvest. Control by using steam sterilization of soil, good **sanitation**, crop rotation, and good **ventilation**. Avoid night temperatures below 60 ℉ and overhead irrigation. Use **preventative** fungicides.

Powdery mildew fungus first **appears** as pale yellow leaf spots. The spots rapidly enlarge to fine cottony growth on the leaf surface. The spots also can occur on the stems and fruit. The fungus causes severe stress on plants as leaves yellow and die. Powdery **spores** produced on the leaf surface spread from plant to plant by air currents. Control through good sanitation, preventative fungicides, and resistant varieties.

Other **diseases** that can **occasionally cause** problems include various **viruses** (cucumber **mosaic** and watermelon mosaic), gray **mold** (Botrytis), **damping-off**, and **crooking**. Crooking is a **physiological disorder** often caused by temperature **extremes**, **excessive** soil moisture, and **nutrition imbalances**. Fruit will become excessively **curved**, reducing its market value.

Vocabulary

delay [dɪˈleɪ] *v.* 延期,推迟
individual [ˌɪndɪˈvɪdʒuəl] *adj.* 单个的
sow [səʊ] *v.* 播种
container [kənˈteɪnə(r)] *n.* 容器
polyethylene [ˌpɒliˈeθəliːn] *n.* 聚乙烯
shade [ʃeɪd] *n.* 阴凉处
emerge [iˈmɜːdʒ] *v.* 出现,露头
transplant [trænsˈplɑːnt] *v.* 移植；移种
apart [əˈpɑːt] *adv.* 相隔；分开
management [ˈmænɪdʒmənt] *n.* 管理
shrink-wrapped *adj.* 用收缩塑料薄膜包装的
moisture [ˈmɔɪstʃə(r)] *n.* 水分
loss [lɒs] *n.* 损失,减少；丢失
expectation [ˌekspekˈteɪʃn] *n.* 期待；预期
mild flavor 清淡的味道
peel [piːl] *v.* 剥皮
gummy stem blight 蔓枯病
fungus [ˈfʌŋgəs] *n.* 真菌
lesion [ˈliːʒn] *n.* 损害,损伤
node [nəʊd] *n.* 节点
pruning wound 修剪伤口
crack [kræk] *v.* 破裂
gummy ooze 黏性软泥
girdle [ˈgɜːdl] *v.* 围绕,环绕
grayish-green 灰绿
water-soaked 被水浸透的
blossom [ˈblɒsəm] *n.* 花
sanitation [ˌsænɪˈteɪʃn] *n.* 卫生
ventilation [ˌventɪˈleɪʃn] *n.* 通风
preventative [prɪˈventətɪv] *adj.* 预防性的
powdery [ˈpaʊdəri] *adj.* 粉(状)的
mildew [ˈmɪldjuː] *n.* 霉
appear [əˈpɪə(r)] *v.* 出现,显现

spore [spɔː(r)] *n.* 孢子
disease [dɪˈziːz] *n.* 疾病
occasionally [əˈkeɪʒnəli] *adv.* 偶尔；偶然
cause [kɔːz] *v.* 引起；导致
virus [ˈvaɪərəs] *n.* 病毒
mosaic [məʊˈzeɪɪk] *n.* 马赛克
mold [məʊld] *n.* 模子
damping-off [ˈdæmpɪŋˈɒf] *n.* 植物幼苗或插枝的腐烂
crooking [ˈkrʊkɪŋ] *n.* 弯曲
physiological [ˌfɪzɪəˈlɒdʒɪkl] *adj.* 生理学的；生理的
disorder [dɪsˈɔːdə(r)] *n.* 混乱，凌乱
extreme [ɪkˈstriːm] *n.* 极端；困境
excessive [ɪkˈsesɪv] *adj.* 过度的，极度的
nutrition [njuˈtrɪʃn] *n.* 营养
imbalance [ɪmˈbæləns] *n.* 不平衡
curved [kɜːvd] *adj.* 弧形的，弯曲的

Notes

1. They also require higher temperatures, which means they are generally grown as a spring or early summer crop. Daytime temperatures should be 80-85 ℉ (nighttime 65-75 ℉). Soil temperatures should be at least 65 ℉.
   它们还需要更高的温度，这意味着它们通常被作为春季或初夏作物。白天的温度应该在 80～85 ℉（夜间 65～75 ℉）。土壤温度至少应为 65 ℉。
2. Plants are best started in individual containers.
   黄瓜最好在单独的容器里种植。
3. After plants have formed at least two true leaves, transplant them to their permanent location in the growing bed.
   植物形成至少两个真正的叶子以后，把它们移植到苗床的固定位置。
4. Seedless European greenhouse cucumbers are distinctly different from traditional field-grown cucumbers.
   无籽的欧洲温室黄瓜明显不同于传统田间种植的黄瓜。
   be different from 与……不同的
5. Excellent selling points include their seedlessness, dark green color, mild flavor, and thin, tender skins that require no peeling.
   突出的卖点包括无籽、深绿色、味道温和，无须剥离的薄而嫩的皮。

6. Gummy stem blight is a fungus that occurs on all above-ground parts of the plant causing extensive damage to leaves, stems, and fruit. Light brown to black lesions occur on leaves, at nodes, and in pruning wounds.
蔓枯病是一种真菌,广泛破坏植物地上部分,会在叶片、茎和果实上蔓延。淡褐色到黑色病变发生在叶片、结节、修剪部位。
7. Powdery mildew fungus first appears as pale yellow leaf spots. The spots rapidly enlarge to fine cottony growth on the leaf surface.
白粉病(白粉病和黄瓜白粉病)真菌首先表现为淡黄色叶斑病。斑点迅速扩大到叶片表面的细棉叶。
8. Other diseases that can occasionally cause problems include various viruses (cucumber mosaic and watermelon mosaic), gray mold (Botrytis), damping-off, and crooking.
其他疾病,有时也会引起包括各种病毒(黄瓜花叶病和西瓜花叶病)、灰霉病、腐烂和弯曲这类问题。

Ⅰ. Translate the paragraphs into Chinese.

Cucumbers generally grow more rapidly than tomatoes and produce earlier. They also require higher temperatures, which means they are generally grown as a spring or early summer crop. Daytime temperatures should be 80-85 ℉ (nighttime 65-75 ℉). Soil temperatures should be at least 65 ℉. Lower temperatures will delay plant growth and fruit development.

Plants are best started in individual containers. As seed are often very expensive, sow one seed per container (1/4 to 1/2 inch deep). Water, cover pots with clear polyethylene, and place in the shade. Plants will emerge in two to three days at 80-85 ℉. Remove plastic coverings when plants emerge and place them in full sun.

After plants have formed at least two true leaves, transplant them to their permanent location in the growing bed. Cucumbers will require 6-8 square feet of space per plant. Plants are generally spaced 2 feet apart in rows 3 to 4 feet apart.

Ⅱ. Complete the following passage with words or expressions given below. Change the forms where necessary.

despite	remote	concept	type	technique	providing
efficiency	method	decreased	solution	defined	traditional
supply	nutrient solution		pesticides		

The 1. ________ of "soilless agriculture" may be new for many of us, but it is important for global agriculture—specifically for 2. ________ food for billions of people at a time when global warming and drought have reduced agricultural 3. ________ . With global warming, the availability of water has 4. ________ , resulting in increasing levels of famine. Soilless agriculture may be the 5. ________ , or at least a part of the solution, in a number of climates. Called hydroponics, this system is also helpful in the prevention of soil damage, the growth of natural vegetables free of 6. ________ and increased productivity. When growing in a soilless system, an accurate 7. ________ of water and nutrients is fundamental. 8. ________ being around for a century, hydroponics was not popularized until the 1930s. Since then, this technology has gone far beyond being used in 9. ________ locations and is now a growing industry that rivals classical soilbased agriculture. Soilless cultivation can be 10. ________ as growing plants without the use of the 11. ________ soil media. This could be done in a 12. ________ with or without the use of artificial media to provide the mechanical support of the plants. There are various 13. ________ of hydroponics growing systems. The main six ones are the reservoir method, the flood and drain method (Ebb and Flow), the drip system, the nutrient film 14. ________ (NFT), the wick system, and the aeroponics 15. ________ .

Ⅲ. Supplemental reading.

Read the following passage and get the main idea. Then discuss with your classmates the differences of fertilizer use mentioned in this passage and that of your area.

Seeking Balance in Fertilizer Use in an Uneven World

Fertilizer use differs from country to country, and from too little to too much. Nitrogen and phosphorus can produce big crops. But they can also pollute water and air.

A recent policy discussion in the journal *Science* compared the nutrient balances of different agriculture systems. Researchers compared the use of fertilizer in three areas that grow maize as a major grain: China, Kenya and the United States.

By two thousand five, they say, farms in northern China produced about the same amount of corn per hectare as farms in the American Midwest. But the Chinese farmers used six times more nitrogen, and produced almost twenty-three times more surplus nitrogen.

Government policies can have an influence. For example, as China sought food security, its policies increased fertilizer use.

The researchers note that farmers in the Midwest used too much fertilizer on their crops through the nineteen seventies. But improved farming methods later increased their yields and, at the same time, made better use of chemical nitrogen fertilizer.

Farms in western Kenya use just over one-tenth as much fertilizer as American farms. Corn harvests remain small. The researchers say farming methods in Sub-Saharan Africa need to improve or else poor quality soil will increase rural poverty. More than two hundred fifty million people do not get enough nutrients from crops to stay healthy.

Nutrient balances in agriculture differ with economic development. Farmers lack enough inputs to maintain soil fertility is parts of many developing countries, especially in Africa south of the Sahara. But countries that are developed or growing quickly often have unnecessary surpluses.

Ammonia gas released by fertilized cropland is a cause of air pollution. The land can also release nitrous oxide, a heat-trapping gas.

Nitrogen runoffs from farms can create large dead zones, like those in the Gulf of Mexico. Algae microorganisms in the water overpopulate because of the surplus nitrogen. The algae take much of the oxygen from the water. Fish and other organisms die.

Laurie Drinkwater at Cornell University in Ithaca, New York, was an author of the report. Professor Drinkwater says farmers need to think about ways to solve some of the causes of nutrient loss from agriculture. She says different countries need different solutions based on location, environment, climate and population needs.

参考译文

温室黄瓜种植

种　植

黄瓜一般比西红柿长得更快，挂果更早。它们还需要更高的温度，这意味着它们通常被作为春季或初夏作物。白天的温度应该在80~85 ℉（夜间65~75 ℉）。土壤温度最低应为65 ℉。更低的温度将延缓植物生长和果实的发育。

植物最好在单独的容器里种植。由于种子价格高，每个容器（1/4至1/2英寸深）里放一粒种子。加水，盖上透明的聚乙烯，放在阴凉处。在80~85 ℉下2~3天植物就长出来了。植物长出后去除覆盖的塑料，放在充足阳光下。

植物形成至少两片真正的叶子以后，把它们移植到苗床的固定位置。每株黄瓜间需要6~8平方英尺的空间。作物一般间距2英尺，行间距3到4英尺。

收 获

在良好的管理下，四个月的时间每株植物可以生产多达20~30磅的水果。欧洲的品种一般在3/4~1磅或12~16英寸大小时收获。果实通常用塑料薄膜收缩包装以防止水分流失软化。在55 ℉与80%~90%的相对湿度下储存水果。

无籽的欧洲温室黄瓜明显不同于传统田间种植的黄瓜。由于消费者对田间种植黄瓜的期盼，温室黄瓜可能需要一些市场推广。优秀的卖点包括无籽、深绿色、味道温和、无须剥离的薄而嫩的皮。

病虫害防治

蔓枯病(didymella bryoniae)是一种真菌，广泛破坏植物地上部分，会在叶片、茎和果实上蔓延。浅褐色到黑色病变发生在叶片、结节、修剪部位。病变的叶片最终变干，脱落。病变的茎破裂后落在土壤上，产生琥珀色的渣，可以导致植物枯死。本病也见于植物花末端灰绿色水渍病变，采摘后在果实上生长。通过土壤蒸汽灭菌，良好的卫生条件，作物轮作和良好的通风来控制。避免夜间温度低于60 ℉和喷灌。使用预防性杀菌剂。

白粉病真菌首先表现为淡黄色叶斑病。斑点迅速扩大到叶片表面。斑点也可以在茎和果实上发生。随着叶子变黄枯死，真菌严重威胁植物。粉状孢子在植物叶表面出现，随着气流在植物间传播。通过良好的卫生，预防性杀菌剂和抗病品种来控制这种病害。

其他疾病，有时也会引起包括各种病毒(黄瓜花叶病和西瓜花叶病)、灰霉病(灰)、腐烂和弯曲这类问题。弯曲是一种生理障碍，通常由极端温度、土壤湿度过大、营养失衡引起。黄瓜会变得过于弯曲，降低其市场价值。

Unit 8
Plant Protection

Lucid waters and lush mountains are invaluable assets.
——绿水青山就是金山银山。

Situational Dialogue:

Talking about Terms of Payment

A: Well, for this grapes valuing \$300,000, we'll pay in installments.

B: That's right. But you should issue a time draft for \$150,000 making payment of the first partial shipment.

A: I know. By the way, when will the draft fall due?

B: On Dec. 31, Please see to it that the draft for the first partial shipment will reach us within the time specified in the contract.

A: Sure. How about the delivery?

B: It will be sent to you in three different lots.

A: By the end of each season, is it?

B: Yes, but I have to remind you that a draft should be issued to us for the next partial shipment after the previous delivery has been made.

A: OK.

Notes

installment: 分散付款
fall due: 到期
see to it: 务必要细致地办好某事
the first partial shipment: 第一批装运(分批装运)
delivery: 交付,递送

Lesson 1 Soil Micro-organisms

There are thousands of millions of every small **organisms** in every ounce of **fertile** soil. Many different types are found but the main groups are:

①**Bacteria**. Bacteria are the smallest types of **single-celled** organisms and can only be seen with a **microscope**. There are many kinds in the soil. Most of them feed on and break down **organic matter**. They obtain energy from the carbohydrates (e. g. sugar, starches, cellulose, etc.) and release carbon dioxide in the process. They also need nitrogen to build cell proteins. If they cannot get this protein from the organic matter they may use other **source** such as the nitrogen applied as fertilizers. When this happens (e. g. where straw is ploughed in) the following crop may suffer from shortage of nitrogen unless extra fertilizer is applied. Some types of bacterica can convert the nitrogen from the air into nitrogen compounds which can be used by plants. Soil bacteria are most active in warm, **damp**, well **aerated** soils which are not **acid**.

②**Fungi**. Fungi are simple types of plants which feed on and break down organic matter. They are mainly responsible for breaking down **lignified (woody) tissue**. They have no chloropghyll or **proper** flowers. The fungi usually found in **arable** soils are very small, but larger types are found in other soils, e. g. **peat**. Fungi can live in acid conditions and in drier conditions than bacteria. (Mushrooms are fungi, and "fairy rings" are produced by fungi.) Sometimes disease-producing fungi develop in some fields, e. g. those causing "take-all" and "**eyespot**" in **cereals**.

③**Actinomycetes**. These are organisms which are **intermediate** between bacteria and fungi and have a similar **effect** on the soil. They need oxygen for growth and are more common in the drier, warmer soils. They are so numerous as bacteria and fungi. Some types can cause plant diseases, e. g. common **scab** in potatoes (worst in light, dry soils).

④**Algae**. Soil algae are very small simple oragnisms which contain chlorophyll and so can build up their bodies by using carbon dioxide from the air and nitrogen from the soil. Algae grow well in fertile damp soils **exposed** to the sun. Algae growing in **swampy** soils can use **dissolved** carbon dioxide from the water and release oxygen. This process is an important source of oxygen for crop plants such as rice. Algae are important in **colonizing bare** soils in the early **stages** of **weathering**.

⑤**Protozoa**. These are very small, single-celled animals. Most of them feed on bacteria and similar small organisms. A few types contain chlorophyll and so can produce carbonhydrates like plants.

The **activities** of the micro-organisms in the soil are rather **complex** and as yet not fully understood, but we do know that they **improve** the **productivity** of the soil. In general, the more fertile the soil, the more organisms there are present.

Vocabulary

micro-organism [ˌmaɪkrəʊˈɔːgənɪzəm] *n.* 微生物
organism [ˈɔːgənɪzəm] *n.* 有机体；微生物
fertile [ˈfɜːtaɪl] *adj.* 富饶的,肥沃的
single-celled 单细胞的
microscope [ˈmaɪkrəskəʊp] *n.* 显微镜
organic [ɔːˈgænɪk] *adj.* [有化] 有机的
matter [ˈmætə] *n.* 物质
source [sɔːs] *n.* 来源
damp [dæmp] *adj.* 潮湿的
aerate [ˈeəreɪt] *v.* 使暴露于空气中
acid [ˈæsɪd] *n.* 酸 *adj.* 酸的
fungus [ˈfʌŋgəs] *n.* 真菌
lignify [ˈlɪgnɪfaɪ] *v.* (使)木质化
woody [ˈwʊdi] *adj.* 木质的
tissue [ˈtɪʃuː] *n.* 组织
proper [ˈprɒpə] *adj.* 适当的
arable [ˈærəbl] *adj.* 适于耕种的
peat [piːt] *n.* 泥炭块
eyespot [ˈaɪˌspɒt] *n.* 眼状斑点
cereal [ˈsɪərɪəl] *n.* 谷类
actinomycete [ˌæktɪnəʊˈmaɪsiːt] *n.* 放射菌类
intermediate [ˌɪntəˈmiːdiət] *n.* [化学] 中间物；媒介
effect [ɪˈfekt] *n.* 影响；效果
scab [skæb] *n.* 痂；疤

alga ['ælgə] *n.* 水藻
expose [ɪk'spəʊz] *v.* 揭露；显示
swampy ['swɒmpi] *adj.* 沼泽的；湿地的
dissolve [dɪ'zɒlv] *v.* 溶解
colonize ['kɒlənaɪz] *v.* 移于殖民地，侵入
bare [beə] *adj.* 空的
stage [steɪdʒ] *n.* 阶段
weather ['weðə(r)] *v.* 经受住
protozoan [ˌprəʊtə'zəʊən] *n.* [无脊椎]原生动物
activity [æk'tɪvɪti] *n.* 活动
complex ['kɒmpleks] *adj.* 复杂的
improve [ɪm'pruːv] *v.* 改善，增进
productivity [ˌprɒdʌk'tɪvəti] *n.* 生产力

Notes

1. thousands of millions of：数以十亿计的，成十亿的
2. feed on (upon)：以……为食物的。如：
   Most of bactcria feed on organic matter.
   大多数细菌以有机物质为食物。
   Sheep feed chiefly on grass.
   羊主要以草为食物。
3. break down：分解，坏掉；(计划)失败；中止，停顿；身体或精神衰弱，如：
   Most of bactcria feed on and break down organic matter.
   大多数细菌以有机物质为食物并分解有机物质。
   His plan was well conceived, but it broke down.
   他的计划设想得很好，但是失败了。
   How did the talks break down?
   谈判怎么中止的？
   He has broken down from illness.
   他由于生病而身体衰弱。
4. “etc.”：拉丁语“et cetera”的简写式。意为“及其他；等等”。相当于英语的 and so on, and so forth, and the rest; and other things。
5. suffer from：受……的害；患；以……为患。多半指疾病对人引起痛苦，但也可以是缺点或其他有害的东西，对主语产生直接不利的影响。如：
   She is suffering from cold. 她患了感冒。
   The following crop may suffer from shortage of nitrogen unless extra fertilizer is applied.

除非另外施肥,否则下一季的庄稼可能因缺氮而受到损害。

6. be responsible for：对……负责。往往不直译,可译为：由于……而……；因为……而……；引起；造成,等等。如：

   Fungi are mainly responsible for breaking down lignified tissue.

   真菌引起木质组织分解。

7. (be) exposed to：暴露于；易受。如：

   Algae grow well in fertile damp soils exposed to the sun.

   藻类在暴露于阳光下的肥沃潮湿的土壤里长得很好。

8. as yet：现在还,目前尚,到目前为止,到当时为止。如：

   As yet we haven't heard from him.

   目前我们还没得到他的消息。

9. (We) do know that they improve the productivity of the soil.

   我们确实知道它们可改善土壤的生产力。

10. In general, the more fertile the soil, the more organisms there are present.

    一般来说,土壤越肥沃,存在的有机体就越多。

    in general：一般来讲,大体说来,通常。如：

    In general, your plan is good.

    大体来说,你的计划很好。

    "the+比较级……(从句),the+比较级(主句)"的结构,第一部分是比较状语从句,第二部分是主句,从句中常省略某些成分,有时主句和从句中的主语和谓语动词都省去,语序也常颠倒。如：

    The sooner, the better. 越快越好。

Exercises

Ⅰ. Fill in the blanks with the words or expressions given below. Change the forms where necessary.

energy　microscope　bacteria　organic matter　nitrogen　carbon dioxide

__________ are the smallest types of single-celled organisms and can only be seen with a __________. There are many kinds in the soil. Most of them feed on and break down __________. They obtain __________ from the carbohydrates (e. g. sugar, starches, cellulose, etc.) and release __________ in the process. They also need __________ to build cell proteins.

Ⅱ. Fill in the blanks with proper words.

Once there lived a rich man 1. __________ wanted to do something for the people of his town. 2. __________ first he wanted to find out whether they deserved his help.

In the centre of the main road into the town, he placed 3. ________ very large stone. Then he 4. ________ (hide) behind a tree and waited. Soon an old man came along with his cow.

"Who put this stone in the centre of the road?" said the old man, but he did not try to remove the stone. Instead, with some difficulty he passed around the stone and continued on his way. 5. ________ man came along and did the same thing; then another came, and another. All of them complained about the stone but not tried to remove 6. ________. Late in the afternoon a young man came along. He saw the stone, 7. ________ (say) to himself: "The night 8. ________ (be) very dark. Some neighbors will come along later in the dark and will fall against the stone."

Then he began to move the stone. He pushed and pulled with all his 9. ________ (strong) to move it. How great was his surprise at last! 10. ________ the stone, he found a bag of money.

Ⅲ. Translate the following sentences into Chinese.

1. There are thousands of millions of every small organisms in every ounce of fertile soil.

2. The activities of the micro-organisms in the soil are rather complex and as yet not fully understood, but we do know that they improve the productivity of the soil.

3. Soil algae are very small simple oragnisms which contain chlorophyll and so can build up their bodies by using carbon dioxide from the air and nitrogen from the soil.

4. Some types of bacterica can convert the nitrogen from the air into nitrogen compounds which can be used by plants.

参考译文

土壤微生物

在每盎司肥沃的土壤中有成千上万个小生物体。已经发现了许多种类型,主要群体包括以下:

①细菌。细菌是最小的单细胞生物,只能在显微镜下看到。在土壤中有许多种细菌,大多数以有机物为食并且分解有机物,它们从碳水化合物(比如糖、淀粉、纤维素等)中获得能

量，同时释放二氧化碳。细菌也需要氮生成细胞蛋白，如果不能从有机物中得到这种蛋白质，它们就会从作为肥料的氮中获得。当这种情况发生时（如把稻草犁入地中）除非另外施肥，否则下一季的庄稼可能因缺氮而受到损害，某些类型的细菌能将空气中的氮转化为植物可用的氮化合物。细菌在温暖、潮湿的非酸性土壤中最活跃。

②真菌。真菌是植物以有机物为食和分解有机物的单一类型，能引起木质组织分解，真菌没有叶绿素也没有花。在耕地土壤中真菌通常是非常小的，在其他土壤，例如泥炭中的真菌可能更大。相比细菌来说，真菌生活在酸性、干燥的环境中（蘑菇是一种真菌，“仙女环”是由真菌产生的），在一些领域有时会培养一些致病真菌，如造成谷物“全蚀病”和“眼状斑点”的真菌。

③放线菌。这是处于细菌和真菌之间的生物体，对土壤有类似的效果，它们需要氧气成长，在干燥、温暖的土壤中更常见。放线菌数量众多，有些种类能引起植物疾病，如土豆疮痂病（光照好、干燥的土壤中最严重）。

④藻类。土壤藻类是含有叶绿素的非常小而简单的一种生物体，通过从空气中获取二氧化碳，从土壤中获取氮气来生长，藻类在暴露于阳光下的肥沃潮湿的土壤里长得很好，在沼泽（浸满水的土壤）中生长的藻类可以利用水中溶解的二氧化碳和释放的氧气，这个过程对于像水稻这样的作物来说是非常重要的氧气来源。在风化的早期阶段，藻类在开垦的裸露的土壤中起了非常重要的作用。

⑤原生动物。这些都是很小的单细胞生物，它们中的大多数以细菌为食，类似于小生物体。有几种类型含有叶绿素，所以像植物一样能产生碳水化合物。

土壤中的微生物活动是相当复杂的，目前为止，我们还无法完全理解，但是我们确实知道它们可改善土壤的生产力。一般来说，土壤越肥沃，存在的有机体就越多。

Lesson 2　The Control of Plant Diseases

Before deciding on control **measures**, it is important to know what is causing the disease. Having **ascertained**, as far as possible, the cause, the **appropriate preventative** or control measure can then be applied.

Crop Rotations

A good crop **rotation** can help to avoid an **accumulation** of the **parasite**. In many cases the organism cannot **exist** except when living on the **host**. If the host plant is not present in the field, in a sense the parasite will be **starved** to death, but it should be remembered that:

Some parasites take years to die, and they may have resting **spores** in the soil waiting for the **susceptible** crop to come along, e. g. **club-root** of the brassicae family.

Some parasites have **alternative** hosts, e. g. funguns causing take-all of wheat is a parasite on some grasses.

Removal of Weed Host

Some parasites use **weeds** as alternative hosts. By controlling the weeds the parasite can be reduced, e. g. **cruciferous** weeds such as **charlock** are hosts to the fungus responsible for club-root.

Both these preventative measures **illustrate** the importance of having a **sound** knowledge of the parasites attacking crops.

Clean Seed

The seed must be from disease. This applies particularly to wheat and **barley** which can carry the fungus causing **loose smut** deeply **embedded** in the grain. Seed should only be used from a disease-free crop.

With potatoes it is essential to obtain clean "seed", free from **virus**. In some districts where the **aphid** is very **prevalent**, potato seed may have to be bought every year.

Resistant Varieties

In plant breeding, although the **breeding** of resistant varieties is better understood, it is not by any means simple. For some years plant breeders concentrated on what if called single or major **gene** resistance. However, with few exceptions, this resistance is overcome by the development of new **races** of the fungus to which the gene is no longer resistant.

Breeding **programs** are now concentrating on **multi-gene** or "field resistance which **means** that a variety has the **characteristics** to **tolerate infection** from a wide **range** of races with little lowering of yield. **Emphasis** is now on **tolerance** rather than resistance".

The Control of Insects

Some insects are **carriers** of parasites causing serious plant disease, e. g. control of the **green-fly** (aphid) in sugar-beet will reduce the **incidence** of virus **yellows**. **Furthermore**, fungi can very often enter through plant **wounds** made by insects.

Remedying Plant Food Deficiencies

In many cases a **deficiency** disease can easily be overcome if the deficient plant food is remedied at an early stage.

The Use of Chemicals

Broadly speaking, chemical control of plant diseases means the use of a fungus killer—a **fungicide**. A fungicide may be applied to the seed, the growing plant, or to the soil. It can be used in the form of a **spray**, **dust** or **gas**. To be effective, it must in no way be harmful to the crop, nor after **suitable precautions** have been taken, to the operator or others, and it must certainly **repay** its cost.

Vocabulary

measure [ˈmeʒə] *n.* 措施
ascertain [ˌæsəˈteɪn] *v.* 确定；查明
appropriate [əˈprəʊprɪət] *adj.* 适当的；恰当的
preventative [prɪˈventətɪv] *n.* 预防法
rotation [rəʊˈteɪʃn] *n.* 轮流
accumulation [əkjuːmjʊˈleɪʃn] *n.* 积聚，累积
parasite [ˈpærəsaɪt] *n.* 寄生虫
exist [ɪɡˈzɪst] *v.* 存在
host [həʊst] *n.* 主人；宿主
starve [stɑːv] *v.* 挨饿
spore [spɔː] *n.* 孢子
susceptible [səˈseptɪbl] *adj.* 易受影响的
club-root [ˈklʌbˌrʊt] *n.* 棒状硬化根
alternative [ɔːlˈtɜːnətɪv] *adj.* 供选择的；选择性的
removal [rɪˈmuːvl] *n.* 移动
weed [wiːd] *n.* 杂草，野草
cruciferous [kruːˈsɪfərəs] *adj.* 十字花科的
charlock [ˈtʃɑːlɒk] *n.* 野芥子
illustrate [ˈɪləstreɪt] *v.* 阐明，举例说明
sound [saʊnd] *adj.* 健全的，全面的
barley [ˈbɑːli] *n.* 大麦
loose [luːs] *adj.* 宽松的
smut [smʌt] *n.* [植保] 黑穗病
loose smut 散黑穗病
embed [ɪmˈbed] *v.* 使嵌入
virus [ˈvaɪrəs] *n.* [病毒] 病毒
aphid [ˈeɪfɪd] *n.* [昆] 蚜虫

prevalent [ˈprevələnt] *adj.* 流行的；普遍的
resistant [rɪˈzɪstənt] *adj.* 抵抗的,反抗的
variety [vəˈraɪəti] *n.* 多样；种类
breed [briːd] *v.* 繁殖
gene [dʒiːn] *n.* [遗]基因,遗传因子
race [reɪs] *n.* 属,种
program [ˈprəʊgræm] *n.* 程序
multi-gene [ˌmʌltɪˈdʒiːn] *n.* 多基因
mean [miːn] *v.* 意味
characteristic [kærəktəˈrɪstɪk] *n.* 特征；特性
tolerate [ˈtɒləreɪt] *v.* 忍受
infection [ɪnˈfekʃn] *n.* 感染；传染
range [reɪndʒ] *n.* 范围；幅度
emphasis [ˈemfəsɪs] *n.* 重点
tolerance [ˈtɒl(ə)r(ə)ns] *n.* 容忍
carrier [ˈkærɪə] *n.* [化学]载体
green-fly [ˈgriːnflaɪ] *n.* 蚜虫
incidence [ˈɪnsɪd(ə)ns] *n.* 发生率；影响
yellow [ˈjeləʊ] *n.* [光]黄色；黄化病
furthermore [fɜːðəˈmɔː] *adv.* 此外；而且
wound [wuːnd] *n.* 创伤,伤口
remedy [ˈremɪdɪ] *v.* 补救；治疗
deficiency [dɪˈfɪʃnsi] *n.* 缺乏
chemical [ˈkemɪkl] *n.* 化学制品,化学药品
fungicide [ˈfʌngɪsaɪd] *n.* 杀真菌剂
spray [spreɪ] *n.* 喷雾
dust [dʌst] *n.* 尘埃
gas [gæs] *n.* 气体
suitable [ˈsuːtəbl] *adj.* 适当的;相配的
precaution [prɪˈkɔːʃn] *n.* 预防措施
repay [riːˈpeɪ] *v.* 偿还;回报

Notes

1. decide on (or upon)：决定
2. as far as possible：尽可能,尽力。如：

Having ascertained, as far as possible, the cause, the appropriate preventative or control measure can then be applied.

尽可能查明原因后,便可采用适当的预防或防治措施。

3. in many cases: 在许多情况下
4. live on (upon): 靠……生活,以……为食。如:

   Most bacteria live on organic matter.

   大多数细菌靠有机物质生活。
5. In many cases the parasite cannot exist except when (it is) living on the host.

   寄生菌除了靠寄主为生以外,在许多情况下就不能生存。
6. in a sense: 在某种意义上

   What you say is right in a sense.

   某种意义上来讲,你说得对。

   If the host plant is not present in the field, in a sense the parasite will be starved to death.

   如果田间没有寄主植物,在某种意义上来说,寄生菌就要饿死。
7. have a... knowledge of: 知道,了解,懂得

   Both these preventative measures illustrate the importance of having a sound knowledge of the parasites attacking crops.

   这两项预防措施说明了充分了解危害作物的寄生菌的重要性。
8. not by any mean=by no means: 绝不

   These goods are by no means satisfactory.

   这些货物绝不令人满意。

   We have ideals, but we are by no means visionaries.

   我们有理想,但我们绝不是空想家。
9. For some years plant breedeers concertrated on what is called single or major gene resistance.

   若干年来,植物育种者集中注意所谓单一的,即主要的基因抗性。
10. Breeding programs are now concentrating on multigene or "field resistance which means that a variety has the characteristics to tolerate infection from a wide range of races with little lowering of yield".

    which 引导的定语从句为非限制性的; 定语从句中 to tolerate...可单独译为一句。

    现在,育种计划正集中在多基因即"田间抗性"上,这就是说,一个品种要具有这样的特性: 能耐各种真菌的感染,同时又不降低产量。
11. a wide range of...: 各种真菌……; a... range of...: 一……行,一……列,一……系列。
12. Emphasis is now on tolerance rather than resistance.

    现在,重点是放在耐性上而不是放在抗性上。
13. Broadly speaking, chemical control of plant diseases means the use of a fungus killer—a fungicide.

    概括地说,植病的化学防治就是使用杀菌剂。

    broadly speaking: 概括地说,笼统说来

14. It must in no way be harmful to the crop.

 它绝不应伤害庄稼。

 in no way：绝不，一点也不

15. It must in no way be harmful to the crop, nor after suitable precautions have been taken, to the operator or others, (= It must in no way be harmful to the crop, nor after suitable precautions have been taken, it must not be harmful to the operator or others either,) and it must certainly repay its cost.

 它绝不应伤害庄稼，在采取适当预防措施后，也不应伤害操作人员或其他人，使其物有所值。

Exercises

Ⅰ. Fill in the blanks with the words given below. Change the forms where necessary.

wheat susceptible starve present exist parasite

A good crop rotation can help to avoid an accumulation of the ________. In many cases the organism cannot ________ except when living on the host. If the host plant is not ________ in the field, in a sense the parasite will be ________ to death, but it should be remembered that: ①Some parasites take years to die, and they may have resting spores in the soil waiting for the ________ crop to come along, e. g. club-root of the brassicae family. ②Some parasites have alternative hosts, e. g. funguns causing take-all of ________ is a parasite on some grasses.

Ⅱ. Fill in each of the blanks with one of the items given.

1. was completed, is completed; is equipped, was equipped

 The hotel, which ________ only last year, ________ with a solarium and sauna.

2. be, is, was, are, were, will be, shall be

 A. Twenty years ________ very long but not long enough to change a Roman nose into a pug.

 B. Three-fourths of the surface of the earth ________ sea.

3. that, which, where, who, whom

 The air then passes to a compressor, ________ it is compressed, and from ________ it is delivered to the combustion chambers.

4. be, is, was, are, were, will be, shall be

 A. He demanded that the dinner ________ paid in US dollars.

 B. He hopes his dream ________ realized after his graduation from college.

5. an, the, these, is, are

 The Japanese ________ industrious people.

6. has, have, itself, themselves, them

   The committee ________ been arguing among ________ all morning.

7. no, not a, neither, nor, none

   A. They proposed several solutions, but ________ seemed to be very satisfactory.

   B. "Can you come on Monday or Tuesday?"

   "I'm afraid ________ day is possible."

8. police, acoustics, nucleus, pajamas, measles, cattle

   A. Singular only: ________

   B. Plural only: ________

Ⅲ. Translate the following sentences into Chinese.

1. Broadly speaking, chemical control of plant diseases means the use of a fungus killer—a fungicide. A fungicide may be applied to the seed, the growing plant, or to the soil.

2. Some insects are carriers of parasites causing serious plant disease, e. g. control of the green-fly (aphid) in sugar-beet will reduce the incidence of virus yellows.

3. The seed must be from disease. This applies particularly to wheat and barley which can carry the fungus causing loose smut deeply embedded in the grain. Seed should only be used from a disease-free crop.

4. Before deciding on control measures it is important to know what is causing the disease. Having ascertained, as far as possible, the cause, the appropriate preventative or control measure can then be applied.

5. A good crop rotation can help to avoid an accumulation of the parasite.

参考译文

植物病害的控制

在采取控制措施之前,先要了解造成疾病的原因。尽可能查明原因后,便可采用适当的预防或防治措施。

作物轮作

作物轮作有助于避免寄生虫的积累。寄生菌除了靠寄主为生以外,在许多情况下就无法生存。如果田间没有寄主植物,在某种意义上来说,寄生菌就要饿死,但要记住:

某些寄生虫需要数年才会死,在土壤中可能会有休眠孢子等待易感染的作物出现,如黑斑病家族中的根肿病。

某些寄生虫有替代寄主,例如造成小麦全蚀病的霉菌就寄生在草上。

清除杂草寄主

一些寄生虫选择杂草作为寄主,通过控制可以减少杂草的寄生虫。比如野芥菜这种十字花科的杂草就是造成根肿病的霉菌的宿体。

这两项预防措施充分说明了了解危害作物的寄生菌的重要性。

干净的种子

种子必须无病,尤其是小麦和大麦,因为它们能携带造成散黑穗病的霉菌,这种病往往根深蒂固,所以只能使用无病种子。

土豆也是必不可少的需要获得干净无病毒的“种子”。在一些地区,蚜虫非常普遍,马铃薯种子可能每年都要买新的。

抗病品种

在植物育种中,虽然抗病品种的育种好理解,但这绝不简单。若干年来,植物育种者集中注意所谓单一的,即主要的基因抗性。然而,除了极少数之外,这种抗病性被新种族的真菌战胜了,这些真菌的基因不再有抗病性。

现在,育种计划正集中在多基因即“田间抗性”上,这就是说,一个品种要具有这样的特性:能耐各种真菌的感染,同时又不降低产量。现在,重点是放在耐性上而不是放在抗性上。

控制昆虫

有些昆虫是造成严重植物病害的寄生虫载体,如控制甜菜中的蚜虫会减少黄化病的发病率。此外,真菌可以通过昆虫造成的植物伤口进入。

加强植物营养

在许多情况下,如果在早期阶段加强植物营养,那么营养缺乏性疾病可以很容易地克服。

使用化学药品

概括地说,植病的化学防治就是使用杀菌剂。杀菌剂可用于种子、植物或土壤。它可以用喷雾、粉尘或气体的形式。它绝不应伤害庄稼,在采取适当预防措施后,也不应伤害操作人员或其他人,使其物有所值。

Lesson 3 The Control of Weed and Plant Diseases

In crop production the control of weeds, diseases and pests is essential to obtain high yields. All three may be controlled by sound farm practices. These include the choice of clean seed and the growing of varieties of crop which can resist disease. They also include careful cultivation, both pre-sowing and post-sowing, and the use of chemicals.

Weeds reduce **crop yields on account of** the fact that they compete with crops for water, soil nutrients and light. They also make harvesting difficult. Most weeds are aggressive and invasive. They grow quickly and spread far, and so are difficult to get rid of. One recommended way of **eradicating** many **persistent** weeds is first to plow up the roots and underground parts of the plant. Then the soil may be cultivated lightly on one or more occasions after the first ploughing.

The principal reason for cultivating the soil is to kill weeds. Weeds may also be killed by means of chemicals which have the **collective** name of **herbicides. Weed-killers** are of two basic types: selective and non-selective. The former remove certain weeds from certain crops. For rice we can spray the herbicide or MCPA over the whole crop at low concentration. The rice will not be affected, but many of the rice weeds will be killed. Non-selective weed killers may be used for removing all vegetation e. g. as brush killers. They must be used extremely carefully for the simple reason that they will cradicate all plants on contact-which include the crop itself. They are usually used before sowing or before the **emergence** of the crop itself.

Plant diseases are caused by organisms which use the crop plant as a "host". These are mainly micro-organisms e. g. fungi, bacteria and viruses. These parasitic micro-organisms live on the food nutrients in the tissue cells of the plants. They frequently kill the host tissues, and either the whole plant or a part of it if damaged and killed.

Micro-organisms are reproduced and spread by minute bodied such as spores, fungi and bacteria. Wind, water, diseased plants, cuttings and tubers, animals, men and insects are some of the means whereby disease is disseminated.

It is very difficult to kill the fungi and bacteria, or to make the virus which is inside the host plant inactive. But the evolution of plant varieties which can resist disease has completely changed the methods of disease control. A number of varieties have been **evolved** and are now available to farmers. So the control of plant diseases has increasingly become a matter of prevention.

Fungi, which attack the aerial parts of the crop, can be controlled by means of fungicide. These are sprayed or dusted on to the plant surfaces. They should be applied before the plant is seriously damaged. Sometimes spray and dust is applied whether disease is present or not. In any case, it is necessary to examine crops frequently for signs of disease.

Soil-borne diseases are much more difficult to control. There are various ways of treating the soil. One way is to use chemicals that easily change into a gas or vapour, which enter the soil and kill the harmful organisms. The soil is covered with a **polythene** sheet and the **volatile** chemical is **injected into** the soil. After about 24 hours the sheet is removed and the soil is allowed to air for a few days before use.

Vocabulary

crop yield 粮食产量
on account of 由于；因为；为了……的缘故
eradicate [ɪˈrædɪkeɪt] *v.* 根除，消灭
persistent [pəˈsɪstənt] *adj.* 坚持的；持久的
collective [kəˈlektɪv] *adj.* 共同的；集合的
herbicide [ˈhɜːbɪsaɪd] *n.* [农药] 除草剂
weed-killer [ˈwiːdˌkɪlə] *n.* 除草剂
emergence [ɪˈmɜːdʒəns] *n.* 出现
evolve [ɪˈvɒlv] *v.* 发展，进化
soil-borne [ˈsɔɪlbɔːn] *adj.* 土传的
polythene [ˈpɒlɪθiːn] *n.* [高分子] 聚乙烯
volatile [ˈvɒlətaɪl] *adj.* [化学] 挥发性的
inject into 把……注入，给……添加

Notes

1. Weeds reduce crop yields on account of the fact that they compete with crops for water, soil nutrients and light.

   杂草降低粮食产量是因为它们与农作物竞争水分、土壤养分和光。

   on account of：由于，因为，为了……的缘故；compete with：与……竞争

2. For rice we can spray the herbicide or MCPA over the whole crop at low concentration.

   对于水稻我们可以喷洒除草剂或对整个作物喷洒低浓度的 MCPA。

   MCPA：二甲四氯

3. Fungi, which attack the aerial parts of the crop, can be controlled by means of fungicide.

   真菌通常会攻击作物的地上部分，所以可以通过杀真菌剂来控制病害。

   by means of：用，依靠，凭借

4. The soil is covered with a polythene sheet and the volatile chemical is injected into the soil.

   土壤覆盖聚乙烯膜，然后注入挥发性化学药品。

   is covered with：被……覆盖；a polythene sheet：聚乙烯膜；injected into：注入

Exercises

Ⅰ. Fill in the blanks with the words or expressions given below. Change the forms where necessary.

plow up	recommend	cultivate	spread	crop yield
aggressive	harvest	compete	plough	

Weeds reduce ________ on account of the fact that they ______ with crops for water, soil nutrients and light. They also make ________ difficult. Most weeds are ________ and invasive.

They grow quickly and ________ far, and so are difficult to get rid of. One ______ way of eradicating many persistent weeds is first to ________ the roots and underground parts of the plant. Then the soil may be _____ lightly on one or more occasions after the first _____.

Ⅱ. Complete each of the sentences with the proper form of the word or phrase given.

1. Since I broke my leg, I ________ (depend) on my niece to see to the daily housework.
2. She ________ (entitle to) a compensation for her damaged car.
3. I guarantee ________ (offer) free repair service within the first three years of your purchase.
4. I suppose there's not much point ________(argue) any further.
5. The more you eat chocolate, ________ (fat) you become.

6. — How long have you been working in your present position?

   — I ________ (work) there for two years by the coming March.

7. Our only request is that this ________ (settle) as soon as possible.

8. I'd rather you ________ (not poke) your nose into her affair. Leave her alone.

Ⅲ. Translate the following sentences into Chinese.

1. The soil is covered with a polythene sheet and the volatile chemical is injected into the soil. After about 24 hours the sheet is removed and the soil is allowed to air for a few days before use.

2. Micro-organisms are reproduced and spread by minute bodied such as spores, fungi and bacteria. Wind, water, diseased plants, cuttings and tubers, animals, men and insects are some of the means whereby disease is disseminated.

3. In crop production the control of weeds, diseases and pests is essential to obtain high yields.

4. Weeds reduce crop yields on account of the fact that they compete with crops for water, soil nutrients and light.

5. The principal reason for cultivating the soil is to kill weeds. Weeds may also be killed by means of chemicals which have the collective name of herbicides. Weed-killers are of two basic types: selective and non-selective.

参考译文

杂草和植物病害的防控

在作物生产中控制杂草、病虫害是获得高产量必不可少的,所有这些都是通过合理的农业实践实现的,这些措施包括选择干净的种子和能够抵抗疾病的作物品种,也包括播种前和播种后的精心培育,此外还有化学药品的使用。

杂草降低粮食产量是因为它们与农作物竞争水分、土壤养分和光,而且也会给收割带来麻烦,大多数杂草都具有侵袭性,它们迅速成长并传播,所以很难除去。要想根除顽固的杂草采取的方法之一就是首先要犁出植物的根和地下部分,然后一次或多次轻微地耕种土壤。

耕种土壤的主要原因是除掉杂草,也可以通过除草剂这种化学药品来除草,除草剂有两种基本类型:选择性和非选择性,前者在某些作物中除掉杂草,对于水稻我们可以喷洒除草剂或对整个作物喷洒低浓度的 MCPA,水稻将不会受到影响,但杂草会被除掉。非选择性除草剂可以用于除掉所有的植被,所以必须非常小心地使用,原因很简单,它们能消灭所有接触到的植物,包括作物本身,这种除草剂通常用于播种前或在作物长出之前。

植物病害是微生物利用作物作为"宿主"造成的,这些微生物主要包括真菌、细菌和病毒。这些寄生的微生物主要靠吸收植物组织细胞的营养生存,它们常常会破坏宿主组织,如果受损,整个植物或部分植物就会死亡。

微生物的复制和传播主要靠孢子、真菌和细菌。风、水、病株、粉屑、块茎、动物、人和昆虫都是一些疾病的传播媒介。

杀死真菌和细菌很难,让宿主植物内的病毒休眠这也很难,但通过植物品种的演变可以抵御疾病,这完全改变了疾病的控制方法,许多品种已经进化,农民可以用来种植,所以植物病害的控制也日益成为如何预防的问题。

真菌通常会攻击作物的地上部分,所以可以通过杀真菌剂来控制病害。杀菌剂可以喷洒在植物表面,在植物严重受损之前就要使用,有时需要判断植物是否有病再使用杀菌剂,在任何情况下,检查农作物是否有患病的迹象非常必要。

土传病害更难控制,有各种方式来处理土壤病害,一种方法是使用很容易就能变成气体或蒸气的化学物质,进入土壤杀死有害生物。土壤覆盖聚乙烯膜,然后注入挥发性化学药品,约 24 小时后把膜去掉,让土壤充分暴露在空气中数天再使用。

Unit 9
Eco-agriculture

> *Geographical indication of agricultural produce can facilitate sustained development of agriculture.*
>
> ——农产品地理标志是促进农业可持续发展的一项重要举措。

Situational Dialogue:

The Quality of the Goods

A: Hello, Ms. Yang. We are here to make a claim with you on the 300 cases of dried fungus.

B: Hello, Mr. Smith. What happened to your goods?

A: Well, we found part of our goods went mouldy.

B: I am sorry to hear that. What is the cause of the mould?

A: On inspection we found that the outside packing was in good condition, but there was something wrong with the inside packing. The plastic bags were full of moisture. It's obvious that not being well-dried is responsible for the mould.

A: We are sorry for our negligence. What is your claim?

B: We demand a replacement of 60 cases. Of course, you have to pay the freight and the inspection fee.

A: I agree to your proposal. I'll arrange shipment as soon as possible. Please accept our apology again.

B: Never mind.

claim: 索赔

case：箱
dried fungus：木耳
mouldy：发霉的
negligence：疏忽；粗心大意

Lesson 1 Green Living：What Is Urban Agriculture?

For hundreds of years, large cities have been almost completely **devoid** of any kind of agriculture. People expected **rural** people to grow the food and bring it into the city **dwellers**, who purchase the fruits of their labors. However, things are **swinging** back around, and people are now interested in what **urban** agriculture is and how to bring it to their cities.

Urban agriculture is when people utilize the space around them to create farms and backyard gardens. Urban agriculture is growing food on a smaller scale than in rural areas. While urban agriculture is sometimes limited in size, there are a number of things that people do to help them expand on the kinds of food they grow and how they grow it.

It is easy enough to define orally what urban agriculture is, but most people want to know examples of how it works. The most famous example of urban agriculture is the South Central Farm. This urban farm in a rundown area of Los Angeles became one of the biggest symbols in the fight to keep food local, and it was the subject of an

award-winning documentary. Examples of urban agriculture that hit closer to home are small-scale city blocks turned into gardens for the **community**.

One of the biggest problems that people have with urban agriculture is that food is grown on a small scale. In order for a city to feed all its **residents**, most of the parks in the area need to **be converted to** food production. Much of the time, gardens pop up in **abandoned** lots in a time of economic uncertainty, only to face problems when the owners of the lots return to claim their **property** when it is able to make money again.

If you are interested in urban agriculture, there are plenty of ways to get involved. Call your local **extension** office, and find out if it offers any chances to help with urban gardens. A lot of **elementary** and middle schools are getting on board with urban gardening and need people to help them. Talk to co-ops and people you meet at farmers markets about a chance to **volunteer** with them.

Every urban garden is different and has different needs. When urban agriculture becomes too large, people find interesting ways to expand. Many places are now putting gardens on rooftops or in the medians of roads.

Vocabulary

devoid [dɪˈvɔɪd] *adj.* 缺乏的
rural [ˈrʊərəl] *adj.* 农村的,乡下的
dweller [ˈdwelə] *n.* 居民,居住者
swing [swɪŋ] *v.* 摇摆
urban [ˈɜːbən] *adj.* 城市的
award-winning *adj.* 获奖的
documentary [dɒkjuˈmentri] *n.* 纪录片
community [kəˈmjuːnəti] *n.* 社区
resident [ˈrezɪdənt] *n.* 居民
be converted to 转变为……,改变为……
abandon [əˈbændən] *v.* 遗弃;放弃
property [ˈprɒpəti] *n.* 财产
extension [ɪkˈstenʃn] *n.* 电话分机
elementary [elɪˈmentri] *adj.* 初级的
volunteer [ˌvɒlənˈtɪə] *n.* 志愿者

Notes

1. rundown area:破坏地区

2. Much of the time, gardens pop up in abandoned lots in a time of economic uncertainty, only to face problems when the owners of the lots return to claim their property when it is able to make money again.

   很多时候,在经济不确定时期,花园都会被废弃,唯一要面对的问题是当花园又能赚钱时,很多业主会回来索要他们的花园。

   pop up:突然出现

3. A lot of elementary and middle schools are getting on board with urban gardening and need people to help them.

   许多中小学都想发展城市园林,而且他们需要人手来帮助他们。

   get on board:入伙

Exercises

Ⅰ. Fill in the blanks with the words given below. Change the forms where necessary.

local	urban	elementary	involved	offer	volunteer

If you are interested in ________ agriculture, there are plenty of ways to get ________. Call your ________ extension office, and find out if it ________ any chances to help with urban gardens. A lot of ________ and middle schools are getting on board with urban gardening and need people to help them. Talk to co-ops and people you meet at farmers markets about a chance to ________ with them.

Ⅱ. Fill in the blanks with the appropriate words.

In the United States, there were 222 people 1. ________ (report) to be billionaires(亿万富翁) in 2003. The 2. ________ of these is Bill Gates, worth at least $41 billion, who made his money 3. ________ starting the company Microsoft. Mr. Gates was only 21 years old 4. __________ he first helped to set up the company in 1976. He was a billionaire 5. ________ the time he was 6. ________, there are still some other people who have made lots of money at even 7. ________ (young) ages. Other young people who have struck it rich include Jackie Coogan and Shirley Temple. 8. ________ of these child actors made over a million dollars 9. ________ (act) in movies before they were 14. But 10. ________ youngest billionaire is Albert von Thurn und Taxis of Germany, who, in 2001, inherited (继承) a billion dollars when he turned 18!

Ⅲ. Translate the following sentences into Chinese.

1. Urban agriculture is when people utilize the space around them to create farms and backyard gardens.

2. It is easy enough to define orally what urban agriculture is, but most people want to know examples of how it works.

3. Much of the time, gardens pop up in abandoned lots in a time of economic uncertainty, only to face problems when the owners of the lots return to claim their property when it is able to make money again.

4. Every urban garden is different and has different needs.

5. Many places are now putting gardens on rooftops or in the medians of roads.

参考译文

绿色生活:什么是城市农业?

几百年来,大城市已经几乎完全没有任何类型的农业了。人们都希望让农民种植食物,然后让城市居民购买他们的劳动果实。然而,风水轮流转,现在人们感兴趣的是都市农业,并且会考虑如何在自己生活的城市发展农业。

都市农业是人们利用周围的空间创造的农场和后花园,都市农业是在比农村地区规模小的地方种植食物,都市农业规模有限,因此,人们所做的大量的事情就是扩大他们种植食品的种类,并且研究怎么种。

什么是都市农业,这很容易定义,但大多数人想知道它具体是怎么操作的。都市农业最著名的例子是南方中心农场。这个农场位于洛杉矶一个破败的地区,它成了保卫城市农业种植的最大标志,也成了一个获奖纪录片的主题。城市农业成功的范例,是把家门口小规模的城市街区变成花园社区。

城市农业最大的一个问题是规模太小。为了能供养所有的居民,该地区的大部分花园需要转换成生产地。很多时候,在经济不确定时期,花园都会被废弃,唯一要面对的问题是当花园又能赚钱时,很多业主会回来索要他们的花园。

如果你对都市农业感兴趣,有很多方式来参与。你可以打电话给当地的推广办事处,但要弄清楚他们是否会给你提供帮助。许多中小学都想发展城市园林,而且他们需要人手来帮助他们。跟合作社或农贸市场的相关人员洽谈,看他们是否能给你提供志愿参与的机会。

每一个都市花园都不一样,而且需求也不一样。当都市农业变得太大时,人们会找其他有趣的方式来发展,现在许多地方都会把花园放在屋顶上或道路中间。

Lesson 2 Introduction to the Fruit-picking Garden

The Nankou Farm, managed by Beijing Capital Agribusiness Group, is located in the south of Nankou Town, Changping. It is one of the first national-level demonstration bases of green food production recognized by the Ministry of Agriculture and also the largest state-owned farm in the suburbs of Beijing. It provides fresh air, beautiful environment and convenient transportation for visitors. Since its **founding** in 1958, it has been awarded more than 30 prizes such as the "Advanced Science and Technology Prize" **issued** by the central government, the Ministry of Agriculture and the **municipal** government. Due to great differences between day and night temperatures and sandy soil, fruit here has a unique flavor. The "Yanguang" Apple is particularly popular among consumers in Beijing and the whole country for its rich juice, crispy flesh, pleasantly sweet and sour flavor and storage stability. Covering an area of nearly 660 **hectares**, the farm **yields** southern fruit such as Taiwan **date**, star fruit, **guava**, **papaya**, **grapefruit** and **loquat**, and dozens of new northern **cultivars** such as strawberry, apple, **nectarine**, **flat peach**, **plum**, **apricot**, **cherry** and pear. The company has developed a base producing high-quality fruit of various kinds, providing services such as **landscape** architecture planning and design, greening programs and technical services. It has established long-term cooperation relationship with many government bodies, institutes, colleges and enterprises and spares no effort to provide opportunities for flower and plant appreciation, team-building activities, popular science education, catering, tourism and farm stay.

The farm is now the first choice of Beijing citizenry to experience **agritourism**, pick fruit and vegetables, adopt trees, and purchase fresh products. In the past 50 years, the company has been dedicated to building its brand image and has enjoyed a good **reputation** in Beijing. Looking into the future. With the support of Beijing Capital Agribusiness Group, it will **persist** in green development and build a high-end brand of modern urban-style agriculture.

Period for fruit picking and tree adoption: December and following January to March for guavas and star fruits, January to March for Taiwan dates, March to May for loquats, November to December and the following January to February for grapefruits, January to May for strawberries, April to September for peaches, May to June for cherries, April to June for apricots, May to September for grapes, June to September for pears and plums, August to November for apples and August to October for dates.

Around the Farm

1. Tourist attractions: China Tank Museum, Baiyanggou Scenic Area, Huyu Scenic Area, Great Wall at Juyong Pass, Fenghuangling Park, and Chici Peace Temple.
2. Hotels: North Great Wall Hotel, Shengli Cultural Garden in Yangfang, Dadu Hotel, and Beijing Home Inn.
3. Health/fitness facilities: North China International Shooting Range and Reignwood Pine Valley Golf Club.
4. Folk custom villages: Yangtaizi Resort, Tianlong Pond Resort.
5. Recommended routes (one day): fruit picking (working on the farm)—enjoying home-style meal (or Yangfang mutton hotpot)—visiting China Tank Museum (or Baiyanggou Scenic Area, Huyu Scenic Area, Great Wall at Juyong Pass, Fenghuangling Park, Chici Peace Temple).
6. Leisure activities: tree adoption, farming, edible wild herbs digging, fruits picking, etc.

Contact Us

Driving:

Drive along Badaling Expressway, take the Changping Science & Technology Park Exit (Exit No. 14), turn left at Shuitun Bridge, drive westwards along Shuinan Road to the end, then turn south and head for about 1,000 meters till the gas station, and then drive westwards for about 800 meters to the fruit-picking garden.

Public transport:

1. Take Bus 20 (Changping) at Longze Light Rail Station to Gecun Village and then walk westwards for about 200 meters.
2. Take Bus 13 (Changping) at Changping Stop directly to the destination.
3. Take Bus 914 in Changping to Gecun Village then walk westwards for about 800 meters.
4. Take Bus 919 at Deshengmen to Nankou and transfer to Bus 20 (Changping) or Bus 914 to Gecun Village, and then walk westwards for about 800 meters.

Address: South Nankou, Changping, Beijing

Postcode: 102202

Telephone: 010-60755064

Fax: 010-89790791
Website: www. bjnknc. com
Email: bjnkncfruit@ 163. com

Vocabulary

founding [ˈfaʊndɪŋ] *n.* 建立,创办
issue [ˈɪʃuː] *v.* 公布,发行;颁发
municipal [mjʊˈnɪsɪp(ə)l] *adj.* 市级的
hectare [ˈhekteə] *n.* 公顷
yield [jiːld] *v.* 出产
date [deɪt] *n.* 枣
guava [ˈgwɑːvə] *n.* 番石榴
papaya [pəˈpaɪə] *n.* 木瓜;[林] 番木瓜树
grapefruit [ˈgreɪpfruːt] *n.* 葡萄柚
loquat [ˈləʊkwɒt] *n.* 枇杷
cultivar [ˈkʌltɪvɑː] *n.* 培育植物
nectarine [ˈnektəriːn] *n.* [园艺]油桃;[园艺]蜜桃
flat peach 蟠桃
plum [plʌm] *n.* 李子;梅子
apricot [ˈeɪprɪkɒt] *n.* 杏,杏子
cherry [ˈtʃeri] *n.* 樱桃
landscape [ˈlændskeɪp] *n.* 风景;风景画
agritourism [ˈægrɪtʊrɪzəm] 农业旅游;观光农业
reputation [repjʊˈteɪʃn] *n.* 名声,名誉
persist [pəˈsɪst] *v.* 坚持;持续

Notes

1. It is one of the first national-level demonstration bases of green food production recognized by the Ministry of Agriculture and also the largest state-owned farm in the suburbs of Beijing.
   这是农业部认定的第一个国家级绿色食品生产示范基地,也是北京郊区最大的国有农场。
   the Ministry of Agriculture: 农业部
2. In the past 50 years, the company has been dedicated to building its brand image and has enjoyed a good reputation in Beijing.
   在过去的 50 年中,公司一直致力于建立自己的品牌形象,在北京享有很好的声誉。

dedicated to：致力于，献身于

Exercises

Ⅰ. Fill in the blanks with the words or expressions given below. Change the forms where necessary.

flavor	environment	award	due to	transportation	issue	fresh

It provides ________ air, beautiful ________ and convenient ________ for visitors. Since its founding in 1958, it has been ________ more than 30 prizes such as the "Advanced Science and Technology Prize" ________ by the central government, the Ministry of Agriculture and the municipal government. ________ great differences between day and night temperatures and sandy soil, fruit here has a unique ________.

Ⅱ. Read each sentence and choose the correct answer.

1. My son got up late this morning. He only had ________ for breakfast.

A. two bread B. two slice of bread

C. two slices of bread D. two slices of breads

2. ________ room is big and bright. They like it very much.

A. Tom and Sam B. Tom's and Sam C. Tom and Sam's D. Tom's and Sam's

3. —Do you know how many ________ a horse has and how many ________ a bee has?

—Of course I know.

A. teeth; feet B. tooth; foot C. foot; teeth D. teeth; foot

4. ________ woman in a purple skirt is Betty's mother.

A. The B. A C. An D. /

5. Now telephones are very popular and they are much ________ than before.

A. cheap B. cheaper C. cheapest D. the cheaper

6. —Hi, Tom. Is your brother as active as you?

—No, he's a quiet boy. He is ________.

A. less outgoing than me B. not so calm as

C. more active than I D. as outgoing as I

7. English ________ in many countries, but Chinese ________ their own language.

A. is spoken; speaks B. speaks; is spoken

C. is spoken; speak D. is spoken; is spoken

8. The young man was often seen ________ by the lake.

A. to draw B. to drawing C. drawing D. drew

9. —So hot in the classroom. Would you mind ________ the window?

—OK, I'll do it right now.

A. not closing　　B. not opening　　C. closing　　D. opening

10. —________ weather! It's raining!

—Bad luck! We can't go climbing today.

A. What bad　　B. What a bad　　C. How bad　　D. How a bad

Ⅲ. Translate the following sentences into Chinese.

1. The Nankou Farm, managed by Beijing Capital Agribusiness Group, is located in the south of Nankou Town, Changping. It is easy enough to define orally what urban agriculture is, but most people want to know examples of how it works.

2. The "Yanguang" Apple is particularly popular among consumers in Beijing and the whole country for its rich juice, crispy flesh, pleasantly sweet and sour flavor and storage stability.

3. It has established long-term cooperation relationship with many government bodies, institutes, colleges and enterprises and spares no effort to provide opportunities for flower and plant appreciation, team-building activities, popular science education, catering, tourism and farm stay.

参考译文

水果采摘园的介绍

南口农场,由北京首都农业集团管理,位于昌平南口镇的南部。这是农业部认定的第一个国家级绿色食品生产示范基地,也是北京郊区最大的国有农场。这里空气清新,环境优美,交通便利。自从1958成立以来,它已获得30多个奖项如由中央政府、农业部和市政府颁发的"科技进步奖"。由于温差大和沙质土壤,这里的水果具有独特的风味。"燕广"苹果由于其多汁、肉脆、酸甜爽口、易储存的特点在北京乃至全国都很受欢迎。农场占地面积近660公顷,主要生产南方水果如台湾椰枣、杨桃、番石榴、木瓜、西柚、枇杷等,和一些北方的新品种如草莓、苹果、油桃、蟠桃、李子、杏子、樱桃、梨。公司开发了一个基地,主要生产各种优质水果并提供服务,如风景园林规划设计、园林绿化方案和技术服务。农场与许多政府机构、研究机构、大学和企业建立了长期的合作关系,不遗余力地提供花卉鉴赏、团队建设活动、科普教育、餐饮、旅游和农家乐。

农场现在是北京市民体验农业旅游、采摘水果和蔬菜、树木养护、购买新鲜产品的第一选择。在过去的50年中,公司一直致力于建立自己的品牌形象,在北京享有很好的声誉。

展望未来,随着北京首都农业集团的支持,它将坚持绿色发展,构建现代城市风貌的一个高端品牌农业。

采摘水果的时间:十二月至次年的一到三月适合采摘番石榴、杨桃,一到三月适合采摘台湾椰枣,三月至五月是枇杷,十一月到十二月至次年一月到二月是葡萄柚,一月到五月是草莓,四月至九月是桃子,五月至六月为樱桃,四月至六月是杏子,五月至九月是葡萄,六月至九月为梨和李子,八月至十一月是苹果,八月至十月是枣。

农场周边

1. 旅游景点:中国坦克博物馆、白杨沟风景区、虎峪风景区、居庸关长城、凤凰岭公园、敕赐和平寺。

2. 酒店:长城大酒店、阳坊胜利文化园、大都酒店、北京如家快捷酒店。

3. 健康/健身设施:中国北方国际射击场和华彬高尔夫俱乐部。

4. 民俗村:长江度假村、天朗塘度假村。

5. 推荐路线(一天):采摘水果(在农场)—享受家常菜(或阳坊涮羊肉)—游览中国坦克博物馆(或白杨沟风景区、虎峪风景区、居庸关长城、凤凰岭公园、敕赐和平寺)。

6. 休闲活动:树木养护、养殖、挖食用野生草药、水果采摘等。

交通方式

自驾:

沿八达岭高速公路,昌平科技园区出口(14号出口),在水屯桥左转,沿水南路向西走到尽头,向南大约1 000米到加油站,然后开车向西约800米到水果采摘园。

公共交通:

1. 乘20路(昌平)从龙泽轻轨站到葛村后向西大约200米。

2. 乘13路(昌平)在常平站直达目的地。

3. 在昌平乘914到葛村再向西步行约800米。

4. 在德胜门坐919路公共汽车到南口换乘20路公交车(昌平)或914到葛村,然后向西步行大约800米。

地址:北京昌平南口镇。

邮编:102202

电话:010-60755064

传真:010-89790791

网站:www. bjnknc. com

电子邮件:bjnkncfruit@ 163. com

Lesson 3 Modern Agricultural Science Demonstration Park at Xiaotangshan

Built in 1998, the Xiaotangshan Modern Agricultural Science Demonstration Park was officially named Changping National Agricultural Park by six ministries including the Ministry of Science and Technology in 2001. It is one of 36 national agricultural parks and the only agricultural science demonstration park in Beijing. In past years, the park has won nine national titles including "National Agricultural Science Park", "Imported Technologies Promotion and Demonstration Base", "Demonstration Site of Industrial Tourism and Agritourism", and "National Education Base of Popular Science". It is also the "Beijing **Municipal** Base of Popular Science Education", and "Beijing Municipal Base of **Patriotic** Education".

The park covers a planned area of 111.4 square kilometers, with a **core** area of 30 square kilometers, which is centered on Daliushu **Roundabout**. The park involves four towns: Xiaotangshan, Xingshou, Cuicun and Baishan. It is the first modern agriculture **project** approved by Beijing Municipal **Commission** for Urban Planning that **integrates** agricultural planning with small **township** construction.

The park is located in Xiaotangshan, close to the north border of the North China Plain. It is crossed by the north-south Litang Road and the east-west Shashun Road and 6th Ring Road. It is 12 kilometers north of the Asian Games Village, 5 kilometers east of Badaling Expressway, and 10 kilometers west of the Capital International Airport. It has fertile soil and is rich in water resources, crossed by eight rivers including the Wenyu, Hulu and Lingou. Its **abundant geothermal** resources spread out over a 100-square kilometer area.

The park has defined its role as "demonstrating modern agricultural science, **fostering** the development of neighboring areas, and promoting agritourism". It is a center of excellence in modern agriculture and technology and in **optimizing** agricultural industry to increase farmers' income. In recent years, seven sections and one garden have been completed: ①a state-of-the-art **sapling cultivation** section; ②an agricultural production section with a 3S positioning system; ③an **aquatic** section **breeding sturgeons** and **Tilapias**; ④a feed and food processing section; ⑤a holiday

resort section featuring hot springs, spa, and gym; ⑥an **organic** fruit and vegetables growing section; ⑦a fresh flowers and gardening section, and a **flora** seeds and seedling cultivation garden (the central garden).

The park is open to tourists all year around. Tourists can learn the latest and advanced agricultural science and visit large greenhouses with various f precious flowers, vegetables, and fruits grown in them, and have the chance to pick fruit and experience the life of modern agriculture.

Since 2005, in order to meet the needs of different visitors, the park has launched activities including students' summer camps and agriculture survey tours, which have been highly praised.

Contact us:

Business hours: 8: 00-17: 00

8: 30-16: 00 (weekends)

Telephone: 010-61791792

Fax: 010-61787260

E-mail: nongyeyuan@ 126. com

Contacts: Miss Liu, Miss Zhao and Miss Huang

Address: North bank of the Wenyu River, Xiaotangshan, Changping, Beijing

Public transport:

Take Bus 643, 803, 985, or 946 to Mafang Stop, and then transfer to Bus 984 or 51.

Driving:

Drive along Litang Road, 6th Ring Road or Badaling Expressway to the destination.

Website: www.nyy.bjchp.gov.cn.

Vocabulary

municipal [mjuːˈnɪsɪpl] *adj.* 市级的；地方自治的
patriotic [ˌpeɪtriˈɒtɪk] *adj.* 爱国的
core [kɔː] *n.* 核心
roundabout [ˈraʊndəbaʊt] *n.* 迂回路线；环状交叉路口
project [ˈprɒdʒekt] *n.* 项目,计划
commission [kəˈmɪʃn] *n.* 委员会
integrate [ˈɪntɪgreɪt] *v.* (使)合并,成为一体
township [ˈtaʊnʃɪp] *n.* 镇区；小镇
abundant [əˈbʌndənt] *adj.* 丰富的

geothermal [dʒiːəʊ'θɜːml] *adj.* [地物]地热的
foster ['fɒstə] *v.* 培养
optimize ['ɒptɪmaɪz] *v.* 优化；使完善
sapling ['sæplɪŋ] *n.* 树苗
cultivation [kʌltɪ'veɪʃn] *n.* 培养；耕种
aquatic [ə'kwætɪk; -'kwɒt-] *adj.* 水生的
breed [briːd] *v.* 生产；培育；使……繁殖
sturgeon ['stɜːdʒən] *n.* 鲟鱼
tilapia [tɪ'læpiə] *n.* 罗非鱼
resort [rɪ'zɔːt] *n.* 度假胜地
organic [ɔː'gænɪk] *adj.* [有化] 有机的
flora ['flɔːrə] *n.* 植物群

Notes

1. In past years, the park has won nine national titles including "National Agricultural Science Park", "Imported Technologies Promotion and Demonstration Base", "Demonstration Site of Industrial Tourism and Agritourism", and "National Education Base of Popular Science".
在过去的几年里，园区赢得了包括"国家农业科技园""进口技术推广示范基地""工农业旅游示范点"和"全国科普教育基地"等九项头衔。
2. The park covers a planned area of 111.4 square kilometers, with a core area of 30 square kilometers, which is centered on Daliushu Roundabout.
园区规划面积111.4平方公里，核心区面积30平方公里，以大柳树环岛为中心。
planned area：规划区域；core area：核心区；be centered on：集中于
3. It is the first modern agriculture project approved by Beijing Municipal Commission for Urban Planning that integrates agricultural planning with small township construction.
它是第一个由北京市委批准的用于城市规划的现代农业项目，此项目把农业规划与小城镇建设紧密地结合了起来。
Beijing Municipal Commission：北京市委；integrate... with：使与……结合
4. The park has defined its role as "demonstrating modern agricultural science, fostering the development of neighboring areas, and promoting agritourism".
园区定位为"展示现代农业科技，促进周边地区发展，促进农业旅游"。
5. Tourists can learn the latest and advanced agricultural science and visit large greenhouses with various f precious flowers, vegetables, and fruits grown in them, and have the chance to pick fruit and experience the life of modern agriculture.
游客可以从中了解到最新最先进的农业科学技术，并且可以参观大型温室，里面有各种各样的名贵花卉、蔬菜、水果，并有机会采摘水果，体验现代农业生活。

Exercises

Ⅰ. Fill in the blanks with the words given below. Change the forms where necessary.

core	integrate	township	cover	involve	agriculture	kilometer

The park ________ a planned area of 111.4 square ________, with a ________ area of 30 square kilometers, which is centered on Daliushu Roundabout. The park ________ four towns: Xiaotangshan, Xingshou, Cuicun and Baishan. It is the first modern ________ project approved by Beijing Municipal Commission for Urban Planning that ________ agricultural planning with small ________ construction.

Ⅱ. Read each sentence and choose the correct answer.

1. We won't give up ________ we should fail ten times.
   A. even if B. since C. whether D. until
2. The teacher spoke loudly ________ the students could hear him clearly.
   A. so as B. that C. so that D. in order to
3. You can have the magazine ________ I finish reading it.
   A. in the moment B. the moment C. the moment as D. in the moment when
4. ________ leaves the room last ought to turn off the lights.
   A. The person B. Anyone C. Who D. Whoever
5. The reason ________ he was late for school was ________ he had to send his mother to a hospital.
   A. that; why B. why; because C. why; that D. that; because
6. Father made a promise ________ I passed the examination he would buy me a bicycle.
   A. that B. if C. whether D. that if
7. ________ you don't like him is none of my business.
   A. What B. Who C. That D. Whether
8. ________ the old man's sons wanted to know was ________ the gold had been hidden.
   A. That; what B. What; where C. What; that D. What; if
9. It is said ________ was all ________ he said.
   A. that; that; that B. what; what; what
   C. that; which; what D. that; that; which
10. He told us ________ he had done. Which of the following is WRONG?
    A. what B. all that C. that D. all what

Ⅲ. Translate the following sentences into Chinese.

1. Built in 1998, the Xiaotangshan Modern Agricultural Science Demonstration Park was officially named Changping National Agricultural Park by six ministries including the

Ministry of Science and Technology in 2001.

2. The park is located in Xiaotangshan, close to the north border of the North China Plain.

3. The park has defined its role as "demonstrating modern agricultural science, fostering the development of neighboring areas, and promoting agritourism".

4. It is a center of excellence in modern agriculture and technology and in optimizing agricultural industry to increase farmers' income.

5. The park is open to tourists all year around. Tourists can learn the latest and advanced agricultural science and visit large greenhouses with various f precious flowers, vegetables, and fruits grown in them, and have the chance to pick fruit and experience the life of modern agriculture.

参考译文

小汤山现代农业科技示范园

小汤山现代农业科技示范园建于1998年,在2001年被包括中国科学技术部在内的六个部委正式命名为昌平国家生态农业园。全国一共有36个农业生态园,这是在北京唯一的一家农业科技示范园。在过去的几年里,园区赢得了包括"国家农业科技园""进口技术推广示范基地""工农业旅游示范点"和"全国科普教育基地"等九项头衔。同时它也是"北京市科普教育基地""北京市爱国主义教育基地"。

园区规划面积111.4平方公里,核心区面积30平方公里,以大柳树环岛为中心,园区包括四个镇:小汤山、兴寿、崔村和白山。它是第一个由北京市委批准的用于城市规划的现代农业项目,此项目把农业规划与小城镇建设紧密地结合了起来。

园区在小汤山,靠近华北平原的北部,境内有南北向的理塘路和东西向的沙顺路和六环路,园区离亚运村12公里、八达岭高速公路5公里、首都国际机场10公里。这里土壤肥沃,水资源十分丰富,有八条河流穿过,包括温榆河、葫芦河和林沟河。其地热资源十分丰富,分布在周围100平方公里的地区。

园区定位为"展示现代农业科技,促进周边地区发展,促进农业旅游"。它是一个集现代

农业技术、优化农业产业、增加农民收入的基地。近年来,已建成了七个区和一个园:①最先进的树苗栽培区;②3S 定位系统的农业生产区;③饲养鲟鱼和罗非鱼的养殖区;④饲料和食品加工区;⑤以温泉、水疗和健身房为主的度假区;⑥有机水果和蔬菜生长区;⑦鲜花和园艺区,以及植物种子和幼苗种植园(中央花园)。

该区一年四季对游客开放。游客可以从中了解到最新最先进的农业科学技术并且可以参观大型温室,里面有各种各样的名贵花卉、蔬菜、水果,并有机会采摘水果、体验现代农业生活。

自 2005 以来,为了满足不同游客的需要,花园推出了一系列的活动,包括学生夏令营和农业调查团,得到了游客们的高度赞扬。

联系方式:

办公时间:8:00—17:00　8:30—16:00(周末)

电话:010-61791792

传真:010-61787260

电子邮件:nongyeyuan@126. com

联系人:刘小姐、赵小姐、黄小姐

地址:北京市昌平县小汤山温榆河北岸

公共交通:

乘坐 643 路、803 路、985 路、946 路公交车道马坊站换乘 984 路或 51 路。

自驾:沿理塘路、六环路、八达岭高速到达目的地。

网址:www.nyy.bjchp.gov.cn.

Unit 10 Applied English

Lesson 1 Resume

依据不同的目的和内容,简历(Resume)有多种写法。通常情况,以下几个部分是必不可少的:1. 个人基本情况(Personal Data);2. 学历(Educational Background);3. 工作经历(Employment History)。其他部分的取舍可根据申请人的实际需要而定。

例如,如果是求职简历,还需要写上:1. 外语技能(Foreign Language Skills);2 取得的资格证书(Certificaites);3. 技能和能力(Qualifications);4. 求职岗位(Applied Position)等。

如果简历是用来申请移民的话,则需要写上:1. 移民经历(Emigration History);2. 打算(Plan);3. 声明(Statement);4. 签名(Signature)及日期(Date)。

Samples:

Resume 1

CLASS 1 NUMBER 012230987

NAME: Cassie

Address:

Nationality: China Telephone: Email:

RELEVANT SKILLS & EXPERIENCE

- Fluent in Chinese, know basic English
- Actively participated in sophomores ERPs and table simulation contest for the post as chief financial officer
- Know the relevant accounting knowledge, at the start of the term had to help

schools charge fees bursary

- Management experience: the class commissary in charge of studies

WORK EXPERIENCE

WENHUA COLLEGE, WUHAN (2009/12/20-2009/12/30)

- Ticket agent: sale train ticket for schoolmate
- Handing flyers for a Japanese training institutions

MINGDIAN COFFEE, WUHAN (2010/1/26-2010/2/25)

- As a waitress to present menus, suggest dishes to customers, etc.

WHITEFISH, MT, USA (2011/6/15-2011/9/15)

- Work for Whitefish Mountain ski & summer resort as a housekeeper

MAJOR SUBJECTS

INTERNATIONAL ECONOMY AND THE TRADE PROGRAM

Concentration courses:

Foundations of International Commerce

Accountant Foundation

International Trade Theory and Practice

International Investment

International Marketing Management

International Commercial Law

International Financial

EDUCATION & QUALIFICATIONS

2009-Present Pass the college entrance examination into Huazhong University of Science & Technology. Glory to join the Chinese communist party.

2006-2009 Study in No. 1 High Cchool.

2003-2006 Study in No. 1 Middle School. Join the Chinese Communist Youth League.

OTHER SKILLS & EXPERIENCE

- COMPUTER SKILLS: Microsoft (Excel, Access, Word, Powerpoint)
- PRIZE: level 4 certificate, level 6 certificate, accounting certificate, merit student

BRIEF INTRODUCTION

Because of my social experience, I have a strong sense of team work. And I become more and more optimistic and responsible for my study and work. What's more, I can keep the passion fresh all the way. I believe that no man is perfect, sometimes I make mistakes, but I know that practice makes perfect. So I will try my best to do and learn something new in the shortest time.

In my spare time, I prefer listening to music and seeing movies. Besides, I also do some exercise by playing badminton, jogging on the playground or jumping rope. So I enjoy a regular life.

Finally, I would really appreciate if I have the opportunity to work there.

Resume 2

1. Nationality Chinese Name Zhao Ming
2. Date of birth: Nov. 15, 1987

(1)Current address: Pomiculture Institude of Agri-Science Academy, 5 Wushan RD. Tianhe District, Guangzhou, 510640, Guangdong, China

(2)Married: No YES√ Name of the spouse: Zhou Li

3. Academic history from elementary education to the latest educaiton

Name of school	Address	Date of admission	Date of graduation
(1)The First Elementary School	Guang Ming Rd., Cheng du City	1996	2002
(2)The Third Middle School	Chengdu City	2002	2008
(3) Agricultural University of Middle China	Wuhan	2008	2012

4. Foreign language studies

Name of school	Address	Date of admission	Date of graduation
__________	__________	__________	__________

5. Employment history (in order of starting dates)

Company/Organization	Address	Starting date	Date of resignation
(1) Institute of Agricultural Academy in Shenzhen	Guangdong	2012	2013
(2) Pomiculture Institude of Agri-Science Academy	Guangzhou	2013	2014
(3) Pomiculture Institute of Agri-Science Academy	Guangzhou	2015	

6. Plan after completion of the program

1)Further academic study 2)Employment 3)Own business 4)Other

(1)Name of the school you wish to advance to: State University of Ohio

Subjects of interest: Plant Protection

(2) Name of the company/Orgnization you plan to join: Fruit Import and

Export Co.

Address of the company/Orgnization: Olympia City, Washington State

Type of business: Fruit Import and Export

7. Others

All of the statements above are of the truth, and I, (name)... have personally written them.

Date: January 20, 2015   Signature: Zhao Xiaoming

Pomiculture Institute of Agri-Science Academy：广东省农业科学院果树研究所

spouse：配偶

Institute of Agricultural Academy in Shenzhen：深圳农业科学研究所

State University of Ohio：美国俄亥俄州立大学

Resume 3

Room 212, Building 343

Tsinghua University, Beijing 100084

(010)62771234 Email: good@ tsinghua.edu.cn

Zheng Yan

Objective

To obtain a challenging position as a software engineer with an emphasis in software design and development.

Education

1997. 9-2000.6 Dept. of Automation, Graduate School of Tsinghua University, M. E. (清华大学研究生院自动化系,获工程硕士学位)

1993. 9-1997.7 Dept. of Automation, Beijing Insititute of Technology, B. E. (北京科技学院自动化专业,获工学士学位)

Academic Main Courses

Mathematics (数学)

Advanced Mathematics Probability and Statistics (高等数学,概率统计)

Linear Algebra (线性代数)

Engineering Mathematics Numerical Algorithm Operational Algorithm(工程数学运算算法的数值算法)

Functional Analysis Linear and Nonlinear Programming (功能分析的线性和非线性规划)

Electronics and Computer(电子计算机)

Circuit Principal Data Structures Digital Electronics (数字电子电路主数据结构)
Artificial Intelligence Computer Local Area Network (人工智能计算机局域网络)

Computer Abilities

Skilled in use of MS Frontpage, Visual Interdev, Distributed Objects, CORBA, C, C++, Project 98, Office 97, Rational Requisite Pro, Process, Pascal, PL/I and SQL software

English Skills

Have a good command of both spoken and written English.
TOEFL: 623; GRE: 2213

Scholarships and Awards

1999. 3 Guanghua First-class Scholarship for graduate (光华学院一等奖学金毕业)
1998. 11 Metal Machining Practice Award (金属加工实践奖)
1997. 4 Academic Progress Award (学术进步奖)

Qualifications

General business knowledge relating to financial, healthcare (金融和医疗保健的一般知识)

Have a passion for the Internet, and an abundance of common sense (对互联网富有激情,有丰富的常识)

Exercises

Ⅰ. Write a resume according to the following information.

王晓华,男,生于1998年6月15日。家住东方市滨海路56号,联系电话为18699112300,电子邮箱 wangxiaohua@163. com。

从2006年9月至2009年7月就读于东方市第一中学。自2010年9月至2013年7月在东方职业技术学院学习,专业为种子生产与经营。曾获得2010年、2011年度奖学金,并于2012年通过计算机考试,获得证书。2013年1月至4月在敦煌种业有限责任公司实习。个人的兴趣爱好是阅读和旅游。

Words for reference:
东方职业技术学院: Dongfang Vocational College
种子生产与经营: Seeds Producing & Marketing
证书: certificate
奖学金: scholarship
实习: internship

Ⅱ. Write an English resume according to your case.

简历常用词汇

1. A Useful Glossary for Personal Data(个人基本情况用语)

name 姓名
alias 别名
pen name 笔名
date of birth 出生日期
birth date 出生日期
born 出生于
birth place 出生地点
age 年龄
native place 籍贯
province 省
city 市
autonomous region 自治区
prefecture 专区
county 县
nationality 国籍
citizenship 国籍
duel citizenship 双重国籍
address 地址
current address 目前地址
present address 目前地址
permanent address 永久地址
postal code 邮政编码
home phone 住宅电话
office phone 办公电话
business phone 办公电话
tel. 电话
sex 性别
male 男
female 女
height 身高
weight 体重
marital status 婚姻状况
family status 家庭状况
married 已婚
single/unmarried 未婚
divorced 离异
separated 分居
number of children 子女人数
none 无
street 街
lane 胡同,巷
road 路
district 区
house number 门牌
health 健康状况
health condition 健康状况
blood type 血型
short-sighted 近视
far-sighted 远视
color-blind 色盲
ID card No. 身份证号码
date of availability 可到职时间
available 可到职
membership 会员,资格
secretary general 秘书长
society 学会
association 协会
research society 研究会

2. A Useful Glossary for Educational Background(教育程度用语)

education 学历
educational background 教育程度
educational history 学历
curriculum 课程
major 主修
minor 副修
educational highlights 课程重点部分
curriculum included 课程包括
specialized courses 专门课程
courses taken 所学课程
courses completed 所学课程
special training 特别训练
social practice 社会实践
refresher course 进修课程
extracurricular activities 课外活动
physical activities 体育活动
recreational activities 娱乐活动
academic activities 学术活动
social activities 社会活动
awards 奖励
scholarship 奖学金
"Three Goods" student 三好学生
excellent League member 优秀团员
excellent leader 优秀干部
student council 学生会
off-job training 脱产培训
in-job training 在职培训
educational system 学制
academic year 学年
semester 学期(美)
term 学期(英)
intelligence quotient 智商
pass 及格
fail 不及格
marks 分数
grades 分数
scores 分数
examination 考试
grade 年级
class 班级
monitor 班长
vice-monitor 副班长
commissary in charge of studies 学习委员
commissary in charge of entertainment 文娱委员
commissary in charge of sports 体育委员
commissary in charge of physical labor 劳动委员
Party branch secretary 党支部书记
League branch secretary 团支部书记
commissary in charge of organization 组织委员
commissary in charge of publicity 宣传委员
degree 学位
post doctorate 博士后
doctor (PhD)博士
master 硕士
bachelor 学士
student 学生
graduate student 研究生
undergraduate 大学生;大学肄业生
senior 大学四年级学生;高中三年级学生
junior 大学三年级学生;高中二年级学生
sophomore 大学二年级学生;高中一年级学生
freshman 大学一年级学生
government-supported student 公费生

commoner 自费生
boarder 寄宿生
classmate 同班同学
schoolmate 同校同学
graduate 毕业生

3. A Useful Glossary for Work Experience(工作经历用语)

accomplish 完成(任务等)
achievements 工作成就,业绩
adapted to 适应于
adept in 善于
administer 管理
advanced worker 先进工作者
analyze 分析
appointed 被任命的
assist 辅助
authorized 委任的;核准的
be promoted to 被提升为
be proposed as 被提名为;被推荐为
behave 表现
breakthrough 惊人的进展,关键问题的解决
break the record 打破纪录
design 设计
develop 开发,发挥
direct 指导
part-time jobs 业余工作
summer jobs 暑期工作
vacation jobs 假期工作
employment experience 工作经历
employment record 工作经历
employment 工作
excellent Party member 优秀党员
guide 指导
implement 完成,实施
job title 职位
lead 领导
mastered 精通的
participate in 参加
position 职位
professional history 职业经历
recommended 被推荐的;被介绍的
representative 代表,代理人
second job 第二职业
specific experience 具体经历
well-trained 训练有素的
work experience 工作经历
work history 工作经历
working model 劳动模范

4. A Useful Glossary for Personal Characters(个人品质用语)

active 主动的,活跃的
adaptable 适应性强的
adroit 灵巧的,机敏的
aggressive 有进取心的
alert 机灵的
ambitious 有雄心壮志的
amiable 和蔼可亲的
amicable 友好的
analytical 善于分析的
apprehensive 有理解力的
aspiring 有志气的,有抱负的
audacious 大胆的,有冒险精神的
capable 有能力的,有才能的
careful 办事仔细的
candid 正直的
charitable 宽厚的

competent 能胜任的
confident 有信心的
conscientious 认真的,自觉的
considerate 体贴的
constructive 建设性的
contemplative 好沉思的
cooperative 有合作精神的
creative 富创造力的
dashing 有一股子冲劲的,有拼搏精神的
dedicated 有奉献精神的
devoted 有献身精神的
dependable 可靠的
disciplined 守纪律的
dutiful 尽职的
earnest 认真的
well-educated 受过良好教育的
efficient 有效率的
energetic 精力充沛的
enthusiastic 充满热情的
expressive 善于表达
faithful 守信的,忠诚的
forceful (性格)坚强的
frank 直率的,真诚的
friendly 友好的
frugal 俭朴的
generous 宽宏大量的
genteel 有教养的
gentle 有礼貌的
hard-working 勤劳的
hearty 精神饱满的
honest 诚实的
hospitable 殷勤的
humorous 幽默的
independent 有主见的
industrious 勤奋的
ingenious 有独创性的
initiative 首创精神
have an inquiring mind 爱动脑筋
intelligent 理解力强的
inventive 有发明才能的,有创造力的
kind-hearted 好心的
knowledgeable 有见识的
learned 精通某门学问的
liberal 心胸宽大的
logical 条理分明的
loyal 忠心耿耿的
methodical 有方法的
modest 谦虚的
motivated 目的明确的
objective 客观的
open-minded 虚心的
orderly 守纪律的
original 有独创性的
painstaking 辛勤的,苦干的,刻苦的
practical 实际的
precise 一丝不苟的
persevering 不屈不挠的
punctual 严守时刻的
purposeful 意志坚强的
qualified 合格的
rational 有理性的
realistic 实事求是的
reasonable 讲道理的
reliable 可信赖的
responsible 负责的
self-conscious 自觉的
selfless 无私的
sensible 明白事理的
sincere 真诚的
smart 精明的
spirited 生气勃勃的
sporting 光明正大的

steady 踏实的
straightforward 老实的
strict 严格的
systematic 有系统的
strong-willed 意志坚强的
sweet-tempered 性情温和的
temperate 稳健的
tireless 孜孜不倦的

5. A Useful Glossary for Other Contents(其他)

objective 目标
career objective 职业目标
employment objective 工作目标
position wanted 希望职位
job objective 工作目标
position applied for 申请职位
position sought 谋求职位
position desired 希望职位
for prospects of promotion 为晋升的前途
for higher responsibility 为更高层次的工作责任
for wider experience 为扩大工作经验
due to close-down of company 由于公司倒闭
due to expiry of employment 由于雇用期满
offered a more challenging opportunity 获得的更有挑战性的工作机会
sought a better job 找到了更好的工作
to look for a more challenging opportunity 找一个更有挑战性的工作机会
to seek a better job 找一份更好的工作

Lesson 2 Application Letter

Application Letter 1

Plot No. 13, Pali Road
Mumbai-345 678
December 26, 2010

Mr. Prakash Ambure

Bank of Mumbai
Andheri, Mumbai 567 889
Dear Mr. Ambure,

This is in reference to your advertisement in the daily newspaper last week regarding a job for the position of a senior accountant. I am highly interested in the position offered and would like to apply my candidature for the same. I am enclosing my CV for your glance.

I am a graduate with accounts and finance as my specialization from a well known school. My academic record has always been excellent and I have been considered among the outstanding students in my school. I have worked for three years as an accountant in a local bank but due to some personal reasons I had to leave the city and shift to Mumbai. I have the qualities that you are looking for in an accountant as mentioned in the advertisement.

I hope you will give me an opportunity to meet you and attend a personal interview and assure you that I would turn an asset to your company.

Thanking you.

Yours faithfully,

Ms. Asha Bhatnagar

Application Letter 2

The following is an example of a letter of application sent with a resume to apply for a job. Your letter should detail your qualifications for the position and the skills you would bring to the employer.

Dear Mr. Gilhooley,

I am writing to apply for the programmer position advertised in the Times Union. As requested, I am enclosing a completed job application, my certification, my resume and three references.

The opportunity presented in this listing is very interesting, and I believe that my strong technical experience and education will make me a very competitive candidate for this position. The key strengths that I possess for success in this position include:

- I have successfully designed, developed, and supported live use applications
- I strive for continued excellence
- I provide exceptional contributions to customer service for all customers

With a BS degree in Computer Programming, I have a full understanding of the full life cycle of a software development project. I also have experience in learning and excelling at new technologies as needed.

Please see my resume for additional information on my experience.

I can be reached anytime via email at john. donaldson@ emailexample. com or my cell phone, 909-555-5555.

Thank you for your time and consideration. I look forward to speaking with you about this employment opportunity.

Sincerely,

Signature (for hard copy letter)

John Donaldson

Reading Materials

英文求职信常用词汇

personnel system 人事制度

personnel management 人事管理

office hour 办公时间

company time 工作时间

endowment insurance 养老保险

medical insurance 医疗保险

unemployment insurance 失业保险

employment injury insurance 工伤保险

maternity insurance 生育保险

housing fund 住房公积金

work hour 工作时间

eight-hour shift 八小时工作制

shift 轮班

morning session 上午班

evening/night shift 小/大夜班

day shift 日班

attendance book 签到本

late book 迟到本

day off 休息日

coffee break 上班中的休息时间

workday 工作日

working condition 工作环境

work permit 工作证

work overtime 加班

holiday rotation 节假日轮流值班

traveling allowance (for official trip) 差旅费

salary 薪水

wage 工资

salary raise 加薪

annual pension 年薪

year-end bonus 年终奖

bonus 奖金

premium 红利

overtime pay 加班费

punch the clock 打卡

time recorder 打卡机

sneak out 开溜

internship 实习

on probation 试用

probation staff 试用人员

agreement of employment 聘书

evaluation of employee 员工考核

employee evaluation form　考核表
merit pay　绩效工资
dock pay　扣薪
unpaid leave　无薪假
before-tax salary　税前薪水
income tax　所得税
take-home pay/after-tax salary　税后
净薪
release pay　遣散费
severance pay　解雇费
salary deduction　罚薪
casual leave　事假
sick leave　病假

Exercises

Ⅰ. Write an application letter according to the following information.

请你以张亮的名义写一封求职信。具体内容如下：

1. 从9月20日的《中国日报》获悉敦煌种业有限责任公司正在招聘对外制种技术员和外贸销售经理,你想参加应聘。
2. 毕业于酒泉职业技术学院,所学专业是作物生产技术。
3. 在校期间通过了大学英语应用能力考试,取得A级资格证书。参加英语口语大赛获得优异成绩。
4. 擅长计算机操作,能熟练运用WPS、Excel、Photoshop等常用软件。
5. 性格开朗,乐于助人。曾担任学生会主席,工作能力较强。
6. 如果求职成功,一定努力工作。

注意信函格式!

Ⅱ. Write an application letter according to the following information.

请根据下列内容写一封求职信。

申请人：王军
写信日期：2019年6月19日
联系地址：海东市东方路450号
联系电话：13670080000
电子邮件：xialei000@163.com
申请职位：销售经理
可开始工作日期：2019年10月15日
个人简历：2017年毕业于东方学院；现在SFG公司工作,担任机械工程师,负责维护机器设备。工作期间,接受过国内与国外的技术培训。
优点：善于沟通,并能很好地与团队成员合作。

Words for reference：
机械工程师：mechanical engineer
维护：maintain

Lesson 3 English Correspondence for International Trade

Reference Templates 1

1. Establishing Business Relation

Dear Sir or Madam,

We are pleased to announce that we intend to intensify our activities in your country. It is our serious and keen interest to realize such a development for our mutual benefit.

We are a trading company involved import and export business throughout the world. We belong to a group of companies established during the turn of 20th century. The attached statement will give you some more information which will surely be helpful to open business relations between us.

Within the activities of the company we have recently established a new department under the management of Mr. Johnson who has long experience in the Trade since 1993. This department is mainly interested in the import of products from your country especially in:

native produce and animal by-products
foodstuffs and chemical industry
raw materials and semi-finished products
chemicals/pharmaceutical raw materials
minerals and essential oils

However, we will be also active in export of chemicals and we invite your inquiries.

We seriously hope that a voluminous and continuous business might be established and please rest assured that we are always doing our utmost to realize a good business relationship with you.

We should be very pleased to be of service to you and hope to submit to you our inquiries soon.

Yours faithfully,

×××

Notes

announce：通知
intensify：加强，加剧
keen：渴望的，热衷的，热心的
statement：财务报告书
pharmaceutical：药学的，药物的，药用的成药，药品
voluminous：广泛的，很多的
utmost：极度的，最大的，最远的；极限，最大限度

2. Online Promotion

Hello,

Beautiful Art is exporter and manufacturer of high quality wooden/iron furniture, gift items, handicraft, antique reproduction at (city), U. K. We also manufacture custom size funiture from solid wood. Other wood species can be used depending on buyers' choice.

Beautiful Art is an ISO 9001: 2000, ISO 14001: 2004, OHSAS 18001: 1999 accredited unit with in-house monthly production of 35~40 containers.

We regularly export furniture to several buyers in Europe and America. The porducts are well accepted there. Kindly visit website www. beautfulartexport. com to view our products.

With regards,

Mooler Smith

Beautiful Art

3. Offers

Dear Sirs,

We thank you for your fax enquiry for both Groundnuts and Walnutmeat CFR Copenhagen dated September 25.

In reply, we offer firm, subject to your reply reaching us on or before Sept. 30 for 250 metric tons of Groundnuts, Handpicked, Shelled and Ungraded at RMB ¥ 2,000 net per metric ton CFR Copenhagen and any other European Main Ports. Shipment to be made within two months after receipt of your order payment by L/C payable by sight draft.

Please note that we have quoted our most favourable price and are unable to

entertain any counter offer.

As you are aware that there has been lately a large demand for the above commodities, such growing demand has doubtlessly resulted in increased prices. However you may avail yourselves of the strengthening market if you will send us an immediate reply.

Yours faithfully,

×××

Notes

groundnut：花生
walnutmeat：核桃仁
shelled：去壳的
entertain：招待、款待
counter offer：还盘
L/C payable：即期信用证

4. Conclusion of Business

Dear Sirs,

We have duly received your Sales Contract No. 5630 covering 50 tons walnutmeat we have booked with you. Enclosed please find the duplicate with our counter signature. Thanks to mutual efforts, we were able to bridge the price gap and put the deal through.

The relative L/C has been established with the Bank of China, London, in your favour. It will reach you in due course.

Regarding further quantities required, we hope you will see your way clear to make us an offer. As an indication, we are prepared to order 80 tons.

Yours faithfully,

×××

Notes

duplicate：副本，复本
mutual：共同的
bridge the gap：弥合差距

Ⅰ. **Translate the letter into English.**

非常高兴接到你公司9月15日番茄酱订单，欢迎你公司成为我公司的客户。

现确认按你方来信列明价格供应番茄酱，并已安排下周由“公主号”轮装出。深信你公司收到货物后，定会感到完全满意。

你公司也许不甚了解我公司的经营范围，现附上目录一份。希望首批订单将建立良好的贸易关系，并带来更多的业务。

Ⅱ. **Write a letter according to the following information.**

假定你是销售部经理王明，请根据下列信息写一封信：

写信日期：2019年6月19日

收信人：Mr. John Brown

内容：

1. 感谢对方订购了你公司的最新产品；
2. 所订购的货物已发出，大约一周后到达；
3. 收到货物后请回复；
4. 希望能继续与对方合作。

注意信函格式！

Words for reference：

deliver：发出　cooperate；cooperation：合作

Reference Templates 2

1. Announcement of Price Increase

Dear ×××,

Due to the increase in raw material costs, we must unfortunately raise the cost of our merchandise to you. We have avoided raising our prices for as long as possible, but we can no longer prolong the inevitable. We have enclosed our new price list for your review which goes into effect on (date). Any order placed between now and (date of increase) will be honored at the lower price. We wish to thank you for your valued account and know that you will understand the necessity for this price increase.

Very truly,

×××

Due to the increase in raw material costs, we must unfortunately raise the cost of our merchandise to you.

由于原材料成本的增加,我们不得不遗憾地向贵公司提高我们的产品价格。

2. Announcement of Price Reduction

Dear ×××,

Rarely do we have the opportunity to inform our customers of such good news. The legislature's tariff ruling which was handed down on May 15th, 1986, has made it possible for our company to reduce our list price for Egyptian cotton. Effective as of June 1, 1986, all full orders received for six-week delivery will be billed as follows:

STOCK	OLD PRICE($)	NEW PRICE ($)
#0134	57.00	51.30
#0135	53.00	47.70
#0136	49.00	44.10

We are very pleased to be able to pass this savings directly on to you. These prices do not include the additional 2% discount that is offered to our customers who pay within the ten-day discount period.

Very truly,

×××

Note

1. Rarely do we have the opportunity to inform our customers of such good news.
   很少有机会能带给我们的客户这样的好消息。
   rarely, never, seldom 是否定的频度副词。不能再和 not 否定词连用,一般只用在句中;若放在句首,则句子要用部分倒装。例句如下:
   Only rarely do I eat in restaurant.
   我极少到餐馆就餐。
   Never have I seen him so angry!
   我从来没有看见过他这样生气!
2. Effective as of June 1, 1986, all full orders received for six week delivery will be billed as follows:

从 1986 年 6 月 1 日起,所有接到的交货期为六周的订单的价格变动如下表:

as of…:从……起,指某事物开始的时间。

3. These prices do not include…

这些价格不包括……

3. Announcement of Special Discount Offer

Dear ×××,

This is to announce our 10% special discount offer that we are making on all orders for the following items for the month of (month) only:

____

____

This 10% discount is available on any order set for delivery from (date) through (date), and is our way of saying thank you for being such a valued customer. We hope you will take advantage of this offer and will send us your purchase order today. We will look forward to hearing from you.

Very truly,

×××

Note

1. This is to announce…:兹宣布……

   This is to notify…:兹通知……

   This is to inform you that/of…:兹告知……

   This is to certify…:兹证明……

2. … is our way of saying thank you for being such a valued customer.

   这是为了感谢贵公司一直以来不吝惠顾。

3. We hope you will take advantage of this offer and will send us your purchase order today.

   希望贵公司能利用这一优惠的报盘今天就发来购买订单。

4. Apology and Replacement of Damaged Goods

Dear ×××,

It was distressing to learn that the chocolate we shipped to your firm last week arrived in bits and pieces. For your request, a new shipment for 30 lbs. left our dock this morning and is scheduled for afternoon delivery to you on May 26th. Please turn over the damaged goods to the driver at the time of delivery. I am sorry that this unfortunate incident occurred and I sincerely appreciate your continued patronage.

Very truly,

×××

1. It was distressing to learn that the chocolate we shipped to your firm last week arrived in bits and pieces.

   得悉上周我方运往贵公司的巧克力在到达时已成碎块了,我们深感苦恼。

   distressing news：令人难过的消息

   a distressing sight：令人伤感的情景

2. I am sorry that this unfortunate incident occurred and I sincerely appreciate your continued patronage.

   对于这次不幸事件的发生,我方深表歉意,并真诚地感谢贵公司继续光顾。

5. Confirmation of Extension of Payment Date

Dear ×××,

This will acknowledge our telephone conversation of this date. As was stated in our letter dated (date), we should be receiving our financing by (date). In view of the above, we would appreciate your extending the date for our payment of the account until (date). This will enable us to make sure that all of the appropriate documents have been prepared. We appreciate your courtesies, past and present. It has been a pleasure doing business with your company and we look forward to years of pleasant associations.

Very truly,

×××

Note

1. This will acknowledge our telephone conversation of this date.

   兹确认我们的这次电话商谈。

   “acknowledge”表示“确认”“承认”时还可以写成:

   sb. acknowledge…, 例如:

   I acknowledge the truth of his statements.

   我承认他说的是事实。

   He acknowledged that the purchase had been a mistake.

   他承认这次采购是个错误之举。

2. In view of the above, we would appreciate your extending the date for our payment of the account until (date). This will enable us to make sure that all of the appropriate documents have been prepared.

   鉴于以上所述,我方将会很感激你方延长我们的付款期限到__日。这样能够确保我方

备好所有适当的文件。

in view of：鉴于；考虑到，例如：

He was pardoned in view of the circumstances.

由于情势的关系，他被特赦了。

6. Denial of Request for Additional Discount

Dear ×××,

This letter is in response to your inquiry regarding our flexibility in the discount rate we offer for early settlement of accounts.

Our established discount is 2% of the total invoiced amount when payment is received within 10 days of delivery. This figure is not one that has been arbitrarily chosen, but is based on cost, overhead and profit. To increase this discount rate for all of our accounts would seriously jeopardize our firm and to increase the rate for an individual account would be both unfair and unethical. I believe that you will find that the 2% discount rate we offer to our customers is standard in the industry.

We consider you a most valued customer and hope that you can appreciate our position in this matter. If we are able to accommodate you in any way that is within our company policy, we will be most happy to do so.

Very truly,

×××

Note

1. This letter is in response to your inquiry regarding…

   这封信是对于你方询问……的一个回复。

2. Our established discount is 2% of the total invoiced amount when payment is received within 10 days of delivery. This figure is not one that has been arbitrarily chosen, but is based on cost, overhead and profit. To increase this discount rate for all of our accounts would seriously jeopardize our firm and to increase the rate for an individual account would be both unfair and unethical. I believe that you will find that the 2% discount rate we offer our customers is standard in the industry.

   如果货款在发货后10天内收到，我们将给予发票金额的2%的既定折扣。这个数字不是随意确定的，而是基于成本、费用及利润计算出来的。如果提高货款折扣的百分比，将会严重影响我们公司的利益，而且如果我们对个别的公司提高折扣的百分比，是不公平也是缺乏职业道德的。我们相信你方会发现我公司给予客户2%的折扣是根据行业的标准确定的。

3. We consider you a most valued customer and hope that you can appreciate our position in this

matter.

我们把你方当作最值得合作的客户,同时,我们也希望在此件事情上你方能给予我们一些理解。

7. New Customer Welcome

Dear ×××,

(name of firm) would like to welcome you as a new customer to our firm. We know that you will be extremely satisfied with our line of products and the service we provide to our customers. You are invited to purchase our merchandise on our regular open account terms (set forth terms). Our credit manager, (name), will be happy to discuss any aspect of our credit policy with you at your convenience. I am enclosing our catalog and price list for your review. I believe that you will find our prices competitive and in keeping with industry trends. Throughout the year we offered our valued customers frequent discounts as an incentive and as a showing of our appreciation.

I do hope you will afford us an opportunity to serve you in the near future.

Very truly,

×××

1. (name of firm) would like to welcome you as a new customer to our firm.

   ________(公司名字)非常欢迎您成为我公司的新客户。

2. I am enclosing our catalog and price list for your review.

   我已经随函寄去了产品目录和价格清单以供您参考。

   "供你参考"相关的写法:

   for your reference

   for your study

   for your perusal

   for your information

3. I believe that you will find our prices competitive and in keeping with industry trends.

   我相信您将会感到我们产品的价格很有竞争力并且紧跟产业的步伐。

   in keeping with:和谐、一致,例句如下:

   A facade had been added, in perfect keeping with (the) original architecture.

   人们给那座建筑加上了一个正面,使其与原建筑风格十分协调。

   out of keeping with:不一致;不协调,例句如下:

   This is out of keeping with the contract.

   这与合同不一致。

4. Throughout the year we offered our valued customers frequent discounts as an incentive and as a showing of our appreciation.

一年来,我们经常给予我们尊贵的客户折扣,以示鼓励和表示感谢。

throughout: 各处,到处; 遍及; 从头到尾; 全部时间

throughout the week: 整个星期

throughout one's life: 毕生,整个一生中

throughout the country: 全国

5. I do hope you will afford us an opportunity to serve you in the near future.

我非常希望在不久的将来您能给我们一个为您服务的机会。

参考译文

国际贸易英文函电

参考模板 1

1. 建立业务关系

敬启者:

特此奉告,我们打算在贵国加强我们的业务活动,并殷切地希望实现这样一个互利的发展。

我们是一家在世界范围内经营进出口的贸易公司。我们隶属于一家20世纪初成立的集团公司。随函寄去的一份报表将向您提供更多的信息,相信它将会对我们之间的贸易关系的开展有所帮助。

我们最近成立了以Johnson先生为领导的新部门,此人从1993年以来就从事此项贸易,经验丰富。该部门主要是从贵国进口下列商品:

- 土畜产品
- 食品和化工产品
- 原材料和半成品
- 化工或医药原料
- 矿产品和香油

当然,我们还将积极出口化工品并欢迎你们询盘。

我们真诚地希望能够建立广泛和持久的贸易,而且请您相信我们将尽最大努力以实现双方友好的业务关系。

我们将很高兴为您服务,并向你提交询盘。

诚挚问候

×××

2. 网络销售

您好：

美艺(Beautiful Art)公司总部位于英国的……市，是一家专门从事高档家具的制造和出口的公司。其主营产品包括木制、铁质家具、礼品、手工艺品、仿古家具，我们还量身定做实木家具。此外，公司还根据顾客需求订制其他材质的家具。

美艺公司已通过 ISO 9001(2000)及 ISO 14001(2004)系列认证，同时还是 OHSAS 18001(职业健康安全管理标准)(1999)认证单位。我们每月能生产 35~40 集装箱的家居产品。

目前，公司的主要客户分布于欧美地区，销售的产品在当地口碑极佳。了解更多产品信息，敬请登录公司网站：www. beautifulartexport. com。

真诚问候

穆勒·史密斯

美艺(Beautiful Art)

3. 报盘

敬启者：

我们已收到你方 9 月 25 日关于花生和核桃仁的哥本哈根的成本加运费传真询价。

此复，我们报实盘以你方时间 9 月 30 日复到为有效，250 公吨花生，手捡、去壳、不分等级，每公吨人民币 2 000 元，成本加运费哥本哈根和其他主要欧洲港口。装运在收到你方用即期信用证支付的订单货款后，两个月内进行。

请你方注意，我们已报了最惠价格因而不能考虑接受任何还盘。

正如你们所知，这种商品目前需求很大，毫无疑问这将导致价格的上涨。然而，如果你们立即答复，你们就可以利用先机。

诚挚问候

×××

4. 达成交易

敬启者：

我们已及时收到了同贵公司签订的 50 吨核桃仁的第 5630 号销货合同。随函寄去我们已会签的副本，请查收。多亏了双方的努力，我们才能弥合价格差距，达成协议。

有关以你方为受益人的信用证已通过伦敦的中国银行开出，并将及时送达你处。

关于进一步所需的数量，我们希望你方能设法给我方报盘。顺便提醒一下，我们打算订购 80 吨。

诚挚问候

×××

参考模板2

1. 涨价通知

敬启者:

由于原材料成本的增加,我们不得不遗憾地向贵公司提高我们的产品价格。

我们已经尽可能地防止价格上涨,但现在我们不能再继续维持原价了,提价已不可避免。

随信附上我们的新价格表供您参阅,该价格表在(日期)内有效。所有自现在起至(涨价的日期)期间订立的订单将可以享受未提高的价格。

感谢贵公司的惠顾,我们深信您会理解我们提价的必要性。

您真诚的,

×××

2. 减价通知

敬启者:

很少有机会能带给我们的客户这样的好消息。立法机构在1986年5月15日颁布的税法,使我公司有可能降低价格表上埃及棉的价格。从1986年6月1日起,所有接到的交货期为六周的订单的价格变动如下表:

货　号	原价(美元)	现价(美元)
#0134	57.00	51.30
#0135	53.00	47.70
#0136	49.00	44.10

我们很高兴能够直接让利给贵方。这些价格不包括我们额外提供给在十天打折期内付款的客户的2%折扣。

您真诚的,

×××

3. 特别打折通知

敬启者:

谨以此敬告,我公司将在(月份)就以下货物的所有订单给予10%的特别折扣:

____

____

这个10%的折扣适用于所有在(日期)到(日期)期间交货的订单,这是为了感谢贵公司一直以来不吝惠顾。

希望贵公司能利用这一优惠的报盘今天就发来购买订单。我们将静候佳音。

您真诚的,

×××

4. 对残损的货物表示道歉并换货

敬启者:

得悉上周我方运往贵公司的巧克力在到达时许多已破碎,我们深感苦恼。应你方的要求,今早一批30磅的新货离开我公司码头,按预定将在5月26日午后抵达你方。请在交付的同时将损坏的货物移交给承运人。

对于这次不幸事件的发生,我方深表歉意,并真诚地感谢贵公司继续光顾。

您真诚的,

×××

5. 延期付款确认

敬启者:

兹确认我们的这次电话商谈。正如我方____日函所述,我方应于____日收到款项。

鉴于以上所述,我方感谢你方能将我方的付款日期延至____日。这样能够确保我方备好所有适当的文件。

我方一直很感谢你方的礼让。很荣幸与贵公司成为贸易伙伴,并期望双方的长期愉快合作。

您真诚的,

×××

6. 拒绝额外打折的要求

敬启者:

这封信是对于你方询问对提前结账是否给予折扣的一个回复。

如果货款在发货后10天内收到,我们将给予发票金额的2%的既定折扣。这个数字不是随意确定的,而是基于成本、费用及利润计算出来的。如果提高货款折扣的百分比,将会严重影响我们公司的利益,而且如果我们对个别的公司提高折扣的百分比,是不公平也是缺乏职业道德的。我们相信你方会发现我公司给予客户2%的折扣是根据行业的标准确定的。

我们把你方当作最值得合作的客户,同时,我们也希望在此件事情上你方能给予我们一些理解。在不违反我公司相关规定的前提下,我们将很愿意满足你方的任何要求。

您真诚的,

×××

7. 欢迎新客户

敬启者:

________(公司名字)非常欢迎您成为我公司的新客户。我们相信您会对我们提供给客户的产品和服务非常满意。

您可以用我们的惯常赊账条件来购买我们的商品(确定的支付方式)。我们的信用经理(名字)将非常高兴在您方便的情况下和您讨论信用政策问题。

我已经随函寄去了产品目录和价格清单以供您参考。我相信您将会感到我们产品的价格很有竞争力并且紧跟产业的步伐。一年来,我们经常给予我们尊贵的客户折扣,以示鼓励和表示感谢。我非常希望在不久的将来您能给我们一个为您服务的机会。

您真诚的,

×××

Lesson 4 Application Form

Application Form for JAS-Certification Organic Farmers/Farmer Groups JAS 认证申请表—有机农业种植(单个种植户/种植联合体)	

We appreciate your interest in our service. To send you an offer and to schedule an on-site inspection, we must have some more details about your project…

能为您提供此种认证服务是我们的荣幸。为了能作出合理报价并合理安排现场检查,请您提供该项目的详细信息……

Type of activity:

种植的模式:

Single farmer or farm unit 单个种植户	

Farmers group 种植联合体	

Complete company name:

完整的公司名称:

Chinese name:

公司名称(中文):

English name:

公司名称(英文):

Competent person:

负责人姓名:

Physical address of holder of certificate (for DHL consignments)

持证者的详细地址(便于 DHL 邮寄)

Street:

街道:

Post Code:

邮编:

Place (City, Province):

地点:

Country:

国家:

Phone:

电话:

Fax:

传真:

E-mail address:

电子邮箱:

Physical address of the farm unit(s)—if different from above

农场地址(如果与以上地址不同)

Name of the unit:

名称:

Place (Village/Community, County, Province):

地点:

Street:

街道:

Country:

国家:

Phone:

电话:

Fax:

传真:

E-mail address:

电子邮箱:

Note: In case of more than one location/address, please copy the format above(!)

注意:如果有多个地址,请一一列出!

Information on the production unit applicable for JAS certification

申请 JAS 认证的相关情况：

Please complete all points (!)

请填写完整！

1)Number of farms：

农场数量：__________

2)Total acreage of the whole farm/farm units：

整个农场的总面积：__________

3)Acreage of the organic farm/farm units：

有机种植的面积：__________

4) Location of the fields (mountain area, industrial zone, in between pastures, etc.)：

有机地块所处的地理位置(山区、工业区、草甸等)：

____

____

____

5)Methods of farm management：

农场管理方法：

a)Do farm diaries exist?

是否有农事记录？ YES□ NO□

b)Do field registers exist?

是否有地块图？ YES□ NO□

c)Do harvest records exist?

是否有收获记录？ YES□ NO□

d)Are purchase/sales receipts kept on file?

是否保存有生产资料采购与农产品销售的发票？ YES□ NO□

6)Number of staffs doing on-farm administration：

农场管理人员数：__________

7)Qualification of the farm manager (Production Process Management Director)：

农场有机生产负责人学历和工作经历：

____

8)Number of staffs responsible for quality control：

负责质量管理的人数：__________

9)Qualification of the quality manager (if applicable)—please describe：

质检经理的学历和工作经历(如果申请)——请描述：

____

____

10)Do you have written Quality Management Standards (QMH)?
是否有书面的质量管理手册(QMH)? YES□ NO□

11)Type and approximate amount of organically produced products:
有机产品的种类和数量:

____

____

12)Structure of the company/project (single unit, farmer's association, cooperation, number of sub-contractors, etc.)— please describe:
公司/项目组织结构(法人单位、农户合作社、协会、委托生产等),请描述:

____

____

Please consider:

请注意:

As soon as we receive sufficient information about your company/project we will send you our offer.
一旦收到贵公司/项目的上述信息,我们会尽快作出报价。

Confirmation:

确认:

■ I understand and accept that my above stated information will be treated by BCS confidentially. Data will be only forwarded to third parties, if I submit a written agreement.
我懂得: BCS 会妥善保管和处理上述信息,只有在收到我的书面指令后,BCS 才会将上述信息提供给第三方。

■ I confirm, that all above mentioned information represents fully and accurately the operation.
我确认: 提供的上述有关有机产品操作的内容准确、全面、真实、可信。

Place:
地点:
Date:
日期:
Signature of the farm/companies representative: ______________________
农场或公司负责人签字:

____

BCS-result of verification BCS:

确认的结果:

The operation/project is...

☐ recommended for JAS certification

☐ not recommended for JAS certification

☐ recommended for JAS certification, in condition that:

Date: ____________, 201 ___

Signature of BCS representative: ______________________________

Lesson 5 Medical Questionnaire

来访者健康问卷

Name
姓名 ______________________

Company Name(if applicable)
公司名称(如果可以告知) ______________________

Contact at Site
联系地址 ______________________

Reason for Visit
来访缘由 ______________________

Please "√" applicable box 请在相应格内打"√"		Yes 是	No 否
1. Have you ever had or been a carrier of: 曾经有或是以下病毒携带者		☐	☐
A food borne disease	一种食物带来的疾病	☐	☐

Typhoid or paratyphoid 伤寒或副伤寒 ☐ ☐
Tuberculosis 肺结核 ☐ ☐
Parasitic infections 寄生性传染病 ☐ ☐

2. Has any close family suffered from any of the above?
你的任何一位家人是否遭受到以上疾病? ☐ ☐

3. Have you or any close contact suffered from any of the following?
你或你周围的人是否遭受以下痛苦? ☐ ☐
Recurring serious diarrhoea or vomiting 复发性严重的腹泻和呕吐 ☐ ☐
Recurring skin trouble 复发性的皮肤病 ☐ ☐
Recurring boils, sties or septic fingers 复发性的疖子、睑腺炎或糜烂性手指
☐ ☐
Recurring discharge from the ears, eyes, gums/mouth 复发性的失聪、失明、龋齿/口中 ☐ ☐

4. Please give details of any other medical problems which may affect your employment as a food handler, for example, recurring gastrointestinal disorder.
请具体给出任何其他医疗问题,这些问题可能会影响你成为一个合格的食品类员工,例如,复发性的肠胃失调。 ☐ ☐

5. Have you been abroad within the last 3 months?
最近三个月内是否出过国? ☐ ☐
If yes, where?
如果有,哪里?

____

____

I declare that all foregoing statements are true and complete to the best of my knowledge and belief.

我声明上述陈述均真实并尽我所知地完成此调查表。

Signed 填写人 ☐

Print Name 打印名 ☐

Date 日期 ☐

Approved by 批准人 ☐

Position 职位 ☐

Appendix 1 Names of Vegetables, Fruits and Crops

algae 藻类
almond 杏仁
core 果核
juice 果汁
skin 果皮
apple 苹果
apricot flesh 杏肉
apricot pit 杏核
apricot 杏子
arbor/tree 乔木
areca nut 槟榔
asparagus 芦笋
aubergine, eggplant 茄子
baby corn 玉米尖
bamboo shoots 竹笋
banana skin 香蕉皮
banana 香蕉
bean curd sheet 豆腐皮
beansprots 绿豆芽
bean 菜豆,豆荚
beechnut 山毛榉坚果
beet 甜菜
bitter orange 酸橙
black bean 黑豆
blackberry 黑莓
broad bean 蚕豆
broccoli 西兰花
brown rice 糙米
brown sugar 红糖,黄糖
cabbage 圆白菜,卷心菜
canned fruit, tinned fruit 罐头水果
carambola 杨桃
carrot 胡萝卜
cashew nuts 腰果
cauliflower 菜花,花椰菜
celery 芹菜
cherry pit 樱桃核
cherry pulp 樱桃肉

cherry 樱桃
chestnut 栗子
chilli 辣椒
Chinese chestnut 板栗
kiwi fruit 猕猴桃
Chinese cabbage 大白菜
Chinese red pepper 花椒
Chinese walnut 山核桃
chive 香葱
coconut milk 椰奶
coconut water 椰子汁
coconut 椰子
coriander 香菜
cornstarch 玉米淀粉
corn 玉米
creamed coconut 椰油
cucumber 黄瓜
cumquat 金橘
curettes 绿皮南瓜
custer sugar 白砂糖(适用于做糕点)
plum 李子,梅子
dark brown sugar 黑糖
date pit 枣核
date 枣
decayed fruit / rotten fruit 烂果
downy pitch 毛桃
dried black mushroom 冬菇
dried chestnuts 干栗子
dried fish 鱼干
shrimp 虾
dry fruit 干果
dwarf bean 四季豆,刀豆
eddoes 小芋头
fig 无花果
filbert 榛子
flat bean 扁豆,豌豆角
flat peach 蟠桃
flesh 果肉
fresh litchi 鲜荔枝
fruit in bags 袋装水果
fruits of the season 应时水果
garlic 蒜
ginger 姜
gingko 白果,银杏
glutinous rice 糯米
golden apple 黄绿苹果
grape fruit 葡萄柚
grape juice 葡萄汁
grape skin 葡萄皮
grapestone 葡萄核
grape 葡萄
green bean 绿豆
green pepper 青椒
greengage 青梅
Hami melon 哈密瓜
hawthorn 山楂
haw 山楂果
honey peach, juicy peach 水蜜桃
honeysuckle 金银花
icing sugar 糖粉
instant noodles 方便面
kernel 仁
kumquat 金橘
leek 韭菜
lemon 柠檬
lettuce 莴苣
lichen 地衣
locust 洋槐、刺槐
loguat 枇杷
long rice 长米
longan pulp 桂圆肉,龙眼肉
longan 桂圆,龙眼
lotus 荷花
maltose 麦芽糖
mandarine 柑橘
mango 芒果

matrimony vine 枸杞
melon 香瓜,甜瓜
monosodium glutamate 味精
moss 苔藓
mushroom 蘑菇
muskmelon 香瓜,甜瓜
mustard, cress 芥菜苗
navel orange 脐橙
noodles 面条
nut meat 坚果仁
nut shell 坚果壳
nut 坚果
okra 秋葵
oleaster 沙枣
olive 橄榄
onion 洋葱
orange peel 柑橘皮
oyster sauce 蚝油
papaya 木瓜
parsley 欧芹
passion fruit 百香果
peach 桃子
pea 豌豆
peony 牡丹
pepper 胡椒
pickled mustard-green 酸菜
pineapple 菠萝
plain flour 纯面粉
polygonum multiflorum 何首乌
pomegranate 石榴
pomelo 柚子
potato 马铃薯
pumpkin 南瓜,西葫芦
radish 萝卜
red bayberry 杨梅
red bean 红豆
red cabbage 紫色包心菜
red chili powder 辣椒粉
red kidney bean 大红豆
red pepper 红椒
rice-noodles 米粉
sago 西米
saltes black bean 豆豉
sesame oil 麻油
sesame paste 芝麻酱
sesame seeds 芝麻
silk noodles 粉丝
soy sauce 酱油
special-grade 特级的
spinach 菠菜
spring onions 葱
star anise 八角
star fruit 杨桃
strawberry 草莓
strong flour 高筋面粉
sweet corn 甜玉米
sweet potato 番薯
taro 大芋头
ternip 白萝卜
tofu 豆腐
tomato 番茄,西红柿
vineger 醋
walnut kernel 核桃仁
walnut 胡桃,核桃
water chestnut 荸荠
flour 面粉

Appendix 2 Vocabulary Related to Plants

absicicic acid 脱落酸 a plant hormone that slows growth

action potential 动作电位 electric signal that propagates along the membrane of a neuron or other excitable cell as a nongraded depolarization

aggregate fruit 聚合果 a fruit derived from a single flower that has more than one carpel

anther 花药 terminal pollen sack of a stamen, where pollen grains containing sperm-producing male gametophytes form

apical meristem 顶端分生组织 undifferentiated tissue located at the tips of roots and shoots

apomixis 单性生殖 ability of some plant species to reproduce asexually through seeds without fertilization by a male gamete

apoplast 非原质体 everything external to the plasma membranes-cell walls, extracellular spaces, and interior of dead cells

apoptosis 细胞凋亡 type of programmed cell death, which is brought about by activation of enzymes that break down many chemical components in the cell

aquaporins 水孔蛋白 transporting proteins that move water across the plasma membrane

auxin 植物生长素,植物激素 natural plant hormone that has a variety of effects, including cell elongation, root formation, secondary growth, and fruit growth

bark 树皮 all tissues external to the vascular cambium

blue-light photoreceptor 蓝色光的光感受器 type of light receptor in plants that initiates a variety of responses, such as phototropism and slowing of hypocotyl elongation

brassinosteroid 油菜素内酯 a steriod hormone in plants that has a variety of effects, including cell elongation, retarding leaf abscission, and promoting xylem differentiation

callus 愈合组织 a mass of dividing, undifferentiated cells growing in culture

carnivorous 肉食 plant that captures animals and digests them

carpel 心皮 ovule producing reproductive organ of a flower, consisting of the stigma, style and ovary

Casparian strip 凯氏带 a belt of suberin that prevents water and minerals from crossing the endodermis via the apoplast

circadian rhythm 生理节奏；昼夜节律 a physiological cycle of about 24 hours that persists even in the absence of external cues

coevolution 共同进化 the joint evolution of two interacting species, each in response to selection imposed by the other

coleoptile 胚芽鞘 the covering of the young shoot of the embryo of a grass seed

colerohiza 胚根鞘 the covering of the young root of the embryo of a grass seed

collenchyma cells 厚角细胞 supporting young parts of the plant shoot

companion cells 伴细胞 nonconducting cells that connect to and help load sugars into sieve-tube elements

complete flower 完全花 having all four basic floral organs (sepals, petals, stamens, carpels)

cork cambium 木栓形成层 eplaces epidermis with periderm

cuticle 外皮 waxy coating on epidermal surface to prevent water loss

cytokinins 细胞激肽类 any of a class of related hormones that regard aging and act in concert with auxin to stimulate cell division, influence the pathway of differentiation, and control apical dominance

day-neutral plant 中性植物 a plant in which flower formation is not controlled by photoperiod or day length

dermal tissue 表皮组织 outer protective covering

determinate growth 有限生长 growth that stops after reaching a certain size

de-etiolation 脱黄化 the changes a plant shoot undergoes in response to sunlight (also known as greening)

dioecious 雌雄异体 having male and female reproductive parts on different individuals of the same species

dormancy 睡眠，冬眠 typified by extremely low metabolic rates and suspension of growth and development

double fertilization 双重受精 when two sperm unite with two cells in the female gametophyte to form the zygote and endosperm

embryo sac 胚囊 formed from growth and division of a megaspore into a multicellular structure that typically has 8 haploid nuclei

endodermis 内皮层 cylinder one cell thick that forms boundary with the vascular cylinder

endosperm 胚乳 nutrient rich tissue that provides nourishment to the developing embryo in angiosperm seeds

epicotyl 上胚轴 embryonic axis above the point of attachment of the cotyledon and below the first pair of miniature leaves

epidermis 表皮 layer of tightly packed on outside of leaf

epiphyte 附生植物 plant that grows on another plant but do not tap into their hosts for sustenance

ethylene 乙烯 a gaseous plant hormone involved in responses to mechanical stress, programmed cell death, leaf abscission, and fruit ripening

etiolation 青白化，苍白化，黄化 plant morphological adaptations for growing in darkness

expansion 扩张物 plant enzyme that breaks the cross-links between cellulose microfibrils and other cell wall constituents, loosening the wall's fabric

flaccid 萎蔫 limp as a result of loss of water

florigen 花激素 a flowering signal, probably a protein that is made in leaves under certain conditions and that travels to the shoot apical meristems, inducing them to switch from vegetative to reproductive growth

fragmentation 分裂 means of asexual reproduction whereby a single parent breaks into parts that regenerate into whole new individuals

fruit 果实 a mature ovary of a flower. protects and often aids in the dispersal.

gene-for-gene recognition 基因对基因的识别 widespread form of plant disease resistance involving recognition of pathogen-derived molecules by the protein products of specific plant disease resistance genes

gibberellin 赤霉素 any of a class of related hormones that stimulate growth in the stem and leaves, trigger the germination of seeds and breaking of bud dormancy, and stimulate fruit development

ground tissue 基本组织 tissue between dermal and vascular tissues that is specialized for storage, photosynthesis, and support

guard cells 保卫细胞 open and close stomata

heat-shock protein 热激蛋白 a protein that helps protect other proteins during heat stress

hypertensive response 高血压反应 a plant's localized defense response to a pathogen, involving death of cells around the site of infection

hypocotyl 下胚轴 embryonic axis below the point of attachment of the cotyledon and above the radicle

imbibition 吸液 the physical adsorption of water into the internal surfaces of structures

incomplete flower 不完全花 lacking sepals, petals, stamens or carpels

indeterminate growth 无限生长 growth that occurs throughout a plant's life

inflorescence 花簇 a group of flowers tightly clustered together

lateral meristem 侧生分生组织 undifferentiated tissue that allows growth in thickness

leaf 叶子 main photosynthetic organ of plants

long-day plants 长日照植物 plant that usually flowers in late spring or early summer when the light period if longer than a critical length

macronutrients 大量要素 C, O, H, N, K, Ca, Mg, P, S

megaspore 大孢子 a spore from a heterozygous plant species that develops into a female gametophyte

meristem identity genes 分生组织特性基因 facilitating the transition from vegetative growth to flowering

mesophyll 叶肉 ground tissue of the leaf that performs photosynthesis

micronutrients 微量元素Cl, Fe, Mn, B, Zn, Cu, Ni, Mo

microspore 小孢子 a spore from a heterosporous plant species that develops into a male gametophyte

multiple fruit 聚合果a fruit derived from an entire inflorescence

mycorrhizae 菌根mutualistic associations between roots and fungi

node 植物的节point at which leaves attach to stem

organ identity genes 器官特性基因facilitating the development of organs

ovule 胚珠 a structure that develops within the ovary of a seed plant and contains the female gametophyte

parasite 寄生植物 plant that grows on another plant and uses the host for water, minerals, etc.

parenchyma cell 薄壁组织细胞performing most of the metabolic functions of the plant

pericycle 中柱鞘outermost layer in vascular cylinder

periderm 胎皮protective tissue that replaces epidermis in older roots and stems

petal 花瓣modified leaf of a flowering plant that are often the colorful parts of a flower, which advertise to insects and other pollinators

phloem 韧皮部transporting sugars from where they are made to where they are needed

photoperiod 光周期physiological response to photoperiod, the relative lengths of night and day

phototropism 趋光性，向光性growth of a plant shoot toward to away from light

phyllotaxy 叶序arrangement of leaves on a stem

phytochrome 光敏色素 type of light receptor in plants that mostly absorbs red light and regulates many plant responses, such as seed germination and shade avoidance

pistil 雌蕊a single carpel or a group of fused carpels

plasmolysis 质壁分离protoplast shrinks from cell wall as a result of loss of water

pollen grain 花粉粒 in seed plants, a structure consisting of the male gametophyte enclosed within a pollen wall

pollen tube 花粉管 forming after germination of pollen grain and functioning in delivery of sperm to ovule

pollination 授粉 the transfer of pollen to the part of a seed plant containing the ovules, a process required for fertilization

radicle 根an embryonic root of a plant

receptacle 容器 the bart of the stamen where the sepals, petals, stamens, and carpels are attached

rhizobacteria 根瘤菌soil bacteria with especially large populations in the rhizosphere

root cap 根冠protecting apical meristem during primary growth

root hairs 根须thin, tubular extension of root epidermal cells that increase surface area

root 根anchoring a vascular plant in the soil, absorbing minerals, and storing carbohydrates

salicylic acid 水杨酸；柳酸 signaling molecule in plants that may be partially responsible for activating systematic acquired resistance to pathogens

scion 接穗，幼枝 the twig grafted into the stock when making a graft

sclerenchyma cells 厚壁组织细胞 specialized for support and strengthening with lignin

second messengers 第二信使 small molecules and ions in the cell that amplify the signal and transfer it from the receptor to other proteins that carry out the response

seed coat 种皮 an tough outer covering of a seed, formed from the outer coat of an ovule

senescence 衰老；老年期 the growth phase in a plant or plant part from full maturity to death

sepal 萼片 modified leaf in an angiosperm that helps enclose and protect a flower bud before it opens

short day plant 短日照植物 plant that flowers in late summer, fall, or winter when the light period is shorter than a critical length

sieve-tube elements 筛管分子 sugar conducting cells that don't have nuclei, ribosomes, vacuoles, and cytoskeletal elements

simple fruits 单果 a fruit derived from a single or several fused carpels

stamen 雄蕊 pollen producing reproductive organ of a flower which consists of an anther and a filament

statolith 平衡石 a specialized plastid that contains dense starch grains and may play a role in detecting gravity

stele 肢柱 collective vascular tissue of a root or stem

stem 茎 organ that raises or separates leaves, exposing them to sunlight

stigma 柱头 a sticky structure at the top of the style that captures pollen

stock 树干 the plant that provides the root system when making a graft

stomata 气孔 pores that allow exchange of carbon dioxide and oxygen

strigolactones 独脚金内酯 plant hormones that inhibits shoot branching, triggers the germination of parasitic plant seeds, and stimulates the association of plant roots with mycorrhizal fungi

style 样式 the stalk of a flowers carpel

sugar sink 糖池 organ that consumes sugar

sugar source 糖源 organ that produces sugar

symplast 共质体 cytosol of all the living cells in a plant

systematic acquired resistance 系统获得性抗性 helping protect healthy tissue from pathogenic invasion

thigmorphogenesis 接触形态反映 response in plants to chronic mechanical stimulation, resulting from increased ethylene production

thigmotropism 向触性 directional growth in response to touch

tracheids 管胞 long, thin cells with tapered ends through which water moves

translocation 输导作用 transport of the products of photosynthesis

transpiration 蒸发 loss of water vapor from plants

tropism 向性；取向；定向 growth response that results in the curvature of whole plant organs toward or away from stimuli due to differential rates of cell elongation

turgid 肿胀 firm as a result of sufficient water

vascular cambium 维管形成层 adding layers of secondary xylem and phloem

vascular tissue 脉管组织 long-distance transport of materials between root and shoot systems

vegetative reproduction 营养生殖 cloning of plants by asexual means

vernalization 种子促熟法 the use of cold treatment to induce a plant to flower

vessel elements 导管分子 wide, short cells to conduct water

xylem 木质部 conducting water and dissolved minerals upward from roots into shoots

Appendix 3　Common Expressions of Agricultural Economy

absentee landlord　外居地主
agricultural commodities market　农业
agricultural products, farm products　农产品
agronomist　农学家
animal husbandry; animal breeding　畜牧业
arable land; tilled land　耕地
arboriculture　树艺学
cash crop　经济作物
cattle farmer　牧场工人
cattle farm　奶牛场
collective farm　集体农场
compensated transfer of land-use rights　土地使用权有偿转让
contract land　承包耕地
cooperative farm　合作农场
country; countryside　农村
countryman　农民,农夫
countrywoman　农民,农妇
cowherd; cowboy　牛仔
crop year; farming year　农事年
dairy farming　乳品业,乳牛业
dairy industry　乳品业
dairy produce; dairy products　乳制品
develpment-oriented agriculture　开发性农业
dry soil　旱田
extensive farming cultivation　粗放经营
fallow　休闲地;休耕地
family side-line business　家庭副业
farm hand　农场短工,农场工人
farm labourers/laborers　农场工人,农业工人
farm　农场

farmer　农户
farming; husbandry　农业
fertile soil　沃土,肥沃的土壤
foodstuffs　食品
fruit grower　果农
fruit growing　果树栽培
grass　草
grassland　草地
gross output value of agriculture　农业总产值
hacienda　庄园
high-yield, high-quality and high-return agriculture　三高农业(高产、高质、高效)
holding　田产
horticulture　园艺
humus　腐殖质
improve the houshold contract responsibility system with remuneration linked to output　完善家庭联产承包责任制
inherit contracts for management of explorative projects　继承开发性生产项目的承包经营权
integrated operations of trade, industry and agriculture　贸工农一体化经营
intensive farming cultivation　集约经营
irrigable land　水浇地
irrigation and water conservancy projects　农田水利基本建设
land reform; agrarian reform　土地改革
land; soil　土壤
landowner　地主,土地拥有者
latifundium; large landed estate　大农场主
lean soil; poor soil　贫瘠土壤
leave the farmland but not one's hometown; shift from farming to other trades within the rural area　离土不离乡
livestock　牲畜
market gardening　商品蔬菜种植业
meadow　草甸
mechanization of farming　农业机械化
mechanized farming　机械化耕作
pasture land　牧场
plot; parcel; lot　地块
ploughman　农夫,犁田者
prairie　大草原

producer　农业工人
promote agriculture by applying scientific and technological advances　科技兴农
ranch　大农场，牧场
rancher　牧场主
rural exodus　农村迁徙
rural population　农村人口
season　季节
settle　佃户
sharecropper　佃农
shepherd　牧人
silviculture　造林学
smallholder; small farmer　小农
system of storing food grain for unforeseen needs　粮食专项储备制度
technology contract responsibility　技术承包制
tenant farmer; leaseholder　土地租用人
to lie fallow　休闲
township enterprise　乡镇企业
vinegrower　葡萄栽植者
vinegrowing; viticulture　葡萄栽培
vintager　采葡萄者
wasteland; barren land　荒地
water and soil conservation　水土保持
water and soil erosion　水土流失

参考答案

Uint 1

Lesson 1

Ⅰ. 1. 谷物 2. 纤维 3. 水果 4. 蔬菜 5. 草料 6. 苗木 7. 林木 8. 皮毛 9. 肥料 10. 繁荣 11. 羊毛 12. 燃料 13. 生物柴油 14. 药品 15. 阿司匹林 16. 磺胺制剂 17. 盘尼西林(青霉素) 18. 原材料 19. 和平 20. 和谐 21. 健康 22. 财富

Ⅱ. DGFAB CIJHE

Ⅲ. 1. meet 2. needs 3. providing 4. enterprise 5. free 6. land 7. integrated 8. animals 9. insects 10. such as

Ⅳ. 1. Literally, agronomy means the "art of managing field". Technically, it means the "science and economics of crop production by management of farm land".

2. Agronomy is defined as "a branch of agricultural science which deals with principles and practices of field crop production and management of soil for higher productivity".

3. Agronomy occupies a pivotal position and is regarded as the mother branch or primary branch.

Ⅴ. Omitted.

Lesson 2

Ⅰ. 1. role 2. enhance 3. explores 4. harvests 5. isolate 6. support 7. promote 8. pursued 9. limited 10. stored

Ⅱ. 1. F 2. T 3. F 4. F 5. F

Ⅲ. 城市林业是把林业由野外和农村引进到人口稠密的经济文化与工商业集中的城市的林业活动。目前大多数城市繁华、喧闹、生态环境恶化,生活在拥挤狭小的城市空间里的人们身体素质下降。发展城市林业可以美化环境、净化空气、减少噪声和调节小气候,改善城市人的生活质量。总体来说,城市林业为解决城市的环境问题提供了一种新的途径。

Ⅳ. Omitted.

Lesson 3

Ⅰ. A. 1. 战国时期 2. 龙骨水车 3. 水力铁锤 4. 春秋时期 5. 水车 6. 管道系统 7. 链泵 8. 土地粗放的作物 9. 花费,以……为代价 10. 作物生产

B. 1. sorghum 2. peanut 3. cotton 4. oilseed 5. ruins 6. seed drill 7. trip hammer 8. the square-pallet chain pump 9. arable land 10. population

Ⅱ. 1. critical 2. concept 3. conservation 4. climate 5. derive 6. generation 7. fertile 8. transport 9. livestock 10. variety

Ⅲ. 1. sufficient 2. import 3. land 4. more 5. save 6. fruits 7. security 8. encourage 9. expense 10. increased

Ⅳ. 1. 虽然只占世界耕地面积的 10%,却生产出了世界 20%的人口所需的食品。

2. 由于中国是一个发展中国家,耕地严重短缺,中国农业一直都是劳动密集型的。

3. 中国是世界上最大的大豆和其他粮食作物的进口国。

4. 尽管对作物生产有严格限制,但是近几年来,中国农产品出口已经大大增加。

Uint 2

Lesson 1

Ⅰ. Asparagus (Stem) Broccoli (Flower) Carrots (Root) Coffee (Seed) Peach (Fruit) Lettuce (Leaf) Celery (Leaf) Cucumber (Fruit) Radish (Root) Green Pepper (Fruit) Sweet Potato (Root) Spinach (Leaf) Orange (Fruit) Peanuts (Seed) Tomatoes (Fruit)

Ⅱ. 1. Plants come from seeds.

2. Seeds wait to germinate (to germinate means to start to grow) until three needs are met: water, correct temperature (warmth), and a good location (such as in soil). Plants need water, warmth, nutrients from the soil, and light to continue to grow.

3. Plant parts do different things for the plant.

Roots act like straws absorbing water and minerals from the soil. Roots help to anchor the plant in the soil so it does not fall over. Roots also store extra food for future use.

Stems do many things. They support the plant. They act like the plant's plumbing system, conducting water and nutrients from the roots and food in the form of glucose from the leaves to other plant parts.

4. A fruit is what a flower becomes after it is pollinated. The seeds for the plant are inside the

fruit.

Ⅲ. **树叶版画**

在树叶的其中一片叶面上涂上颜料。若颜料涂在叶脉较突出的树叶背面,版画线条会更清晰。

如果要印在T恤衫上,那就在衣物下或衣物之间放上报纸。把涂好颜色的叶面放在衣物上(棉质织物效果好)或纸上,然后在叶面的各个部位均匀用力压。这种方法叫作压印。只要稍加练习就能找到压树叶的力度和颜料的涂抹用量。

圆木棒有时能使这个过程更简单。衣柜内横木切割成12英寸长就是一套便宜的圆木棒工具。也可以使用印制技巧。把印过的树叶放下,朝上的一面涂上颜料,再将印上树叶版画的材料放在树叶上面,一次性紧紧压住材料和树叶。

Ⅳ. Omitted.

Lesson 2

Ⅰ. 1. F. A plant's stem supports the plant and transports water, minerals and glucose.

2. F. Most roots are in the soil, which contains water and minerals.

3. T. Some plants have only one main root, with many tiny roots growing from it.

4. T. Roots store some of this glucose in the form of starch.

5. F. Strawberries, grasses and weeds are herbaceous plants.

6. F. A woody plant may lose its leaves for part of the year, but its stem remains alive. Examples of woody plants are strawberries, grasses and weeds.

Ⅱ. 1. That is the job of the plant's roots.

2. The plant's leaves use water to make food in the form of glucose. But the function leaves all have in common is that they make food for plants.

3. A plant's stem supports the plant and transports water, minerals, and glucose.

4. The flowers are held in place by their underground roots.

5. There are two types of roots. Some plants have only one main roots, with many tiny roots growing from it. This type of root is called taproot. Dandelions have taproots. Other plants have many smaller roots that spread out like tiny arms.

Ⅲ. 中国农业必须走资源节约型、生产清洁型、环境友好型、质量效益型的低碳农业发展道路。因此,在技术上,要大力开展一系列关键技术体系的研发及其推广应用。比如:资源节约型技术,包括节能技术、节地技术、节水技术、节肥技术、节药技术、节种技术、节料技术、省工技术等,以及农用化学品的减量化使用及其代替技术、高光效和高碳汇的植物新品种培育技术、土壤碳汇技术、清洁能源技术、环境友好型清洁生产技术和废弃物无害化处理与资源化利用技术等。

Lesson 3

Ⅰ. 1. H 2. F 3. G 4. E 5. A 6. B 7. C 8. D 9. J 10. I

Ⅱ. 1. A plant needs warmth, light, water, carbon dioxide and about a dozen other chemical elements which it can obtain from the soil.

2. Growth is best between 16 ℃(60 ℉) and 27 ℃(80 ℉).

3. Without light, plants cannot produce carbohydrates and will soon die.

4. Water is required in extravagant amounts for the process of transpiration. Water carries nutrients from the soil into and through the plant and also carries the products of photosynthesis from the leaves to wherever they are needed.

5. Plants need carbon dioxide for photosynthesis.

6. In order that a plant may build up its cell structure and function as a food factory, many simple chemical substances are needed.

Ⅲ. 随着人类社会的经济发展与文明进步,人们对森林的认识有了重大转变,社会对林业的需求也发生了很大的变化,森林维护与改善生态环境的功能在世界范围内得到了广泛的重视和关注。1992年联合国召开的环境与发展大会以"赋予林业以首要地位"为最好级别的政治承诺,并特别强调"在世界最高级会议要解决的问题中,没有任何问题比林业更重要了",将林业问题提高到前所未有的高度,这是世界文明发展史上一个重要的里程碑。

Unit 3

Lesson 1

Ⅰ. 1. 传粉,授粉 2. 花粉 3. 雄蕊 4. 雌蕊 5. 种子 6. 幼苗 7. 胚珠 8. 自花授粉 9. 异花授粉 10. 花蜜 11. 花瓣

Ⅱ. 1. starts 2. creation 3. male 4. top 5. produces 6. transfer 7. rely 8. pollinate 9. brightly 10. attract

Ⅲ. Pollinators are not all the same. Different pollinators feed on different plants and therefore pollinate different plants. Some common pollinators are bees, butterflies, birds, and moths. But that's not all. Other pollinators include flies, beetles, and bats.

Bee

Butterfly

Bird

Moth

Ⅳ. 1. Pollination is the transfer of pollen from a stamen to a pistil. Pollination starts the production of seeds.

2. Well, it all begins in the flower. Flowering plants have several different parts that are important in pollination. Flowers have male parts called stamens that produce a sticky

powder called pollen. Flowers also have a female part called the pistil. The top of the pistil is called the stigma, and is often sticky. Seeds are made at the base of the pistil, in the ovule. To be pollinated, pollen must be moved from a stamen to the stigma.

3. There are two types of pollination. Self-pollination and cross-pollination. When pollen from a plant's stamen is transferred to that same plant's stigma, it is called self-pollination. When pollen from a plant's stamen is transferred to a different plant's stigma, it is called cross-pollination. Cross-pollination produces stronger plants. The plants must be of the same species.
4. Pollination occurs in several ways. People can transfer pollen from one flower to another, but most plants are pollinated without any help from people. Usually plants rely on animals or the wind to pollinate them.
5. Plants such as mosses and ferns reproduce by spores. Cone-bearing plants, like pine or spruce trees for example, reproduce by means of pollen that is produced by a male cone and travels by wind to a female cone of the same species. The seeds then develop in the female cone.

Lesson 2

Ⅰ. 1. ensure 2. survive 3. Similar 4. energy 5. exit 6. replace 7. gain 8. summarize 9. happen 10. reverse

Ⅱ. 1. 叶绿体是大多数植物细胞中的一个微小结构。它含有一种叫叶绿素的物质,叶绿素使植物的叶子和其他部分呈现绿色。
2. 光合作用中,叶绿素利用来自太阳光、空气中的二氧化碳和水形成葡萄糖和氧气。
3. 在蒸腾作用中,损失的水量取决于空气的温度、风、空气中的含水量和土壤。
4. 如果蒸腾作用时损失的水比根吸收的水更多,植物就会枯萎甚至死亡。
5. 光合作用以二氧化碳、水和能量生成葡萄糖和氧气。细胞呼吸则用葡萄糖和氧气生成二氧化碳、水和能量。
6. 动物吸入空气中的氧气。为了使能量为其所用,植物吸收氧气改变食物中的能量。在光合作用过程中植物利用能量生产更多的食物和氧气。二氧化碳—氧气循环确保每一个过程都有足够的氧气和二氧化碳。

Lesson 3

Ⅰ. 1. T 2. F 3. T 4. T 5. F 6. F 7. F

Ⅱ. 1. Plants need several things to make their own food. They need: chlorophyll, a green pigment or color, found in the leaves of plants that helps the plant make food; light (either natural sunlight or artificial light, like from a light bulb); carbon dioxide (CO_2) (a gas found in the air; one of the gases people and animals breathe out when they exhale); water (which the plant collects through its roots); nutrients and minerals

(which the plant collects from the soil through its roots).

2. Plants make food in their leaves. The leaves contain a pigment called chlorophyll, which colors the leaves green. Chlorophyll can make food the plant can use from carbon dioxide, water, nutrients, and energy from sunlight. This process is called photosynthesis.

3. People plant some seeds, but most plants don't rely on people. Plants rely on animals and wind and water to help scatter their seeds. And some plants disperse their seeds in other ways... Some plants have unique ways to disperse their seeds. Several kinds of plants "shoot" seeds out of pods. The seeds can travel quite a few feet from the plant this way.

"Dispersal" means scattering or distribution of something.

Ⅲ. 1. 光合作用过程中,植物释放氧气到空气中。人和动物都需要氧气来呼吸。

2. 首先,一些植物,像右边的毛刺,有可以缠在动物毛皮或羽毛上的倒钩或其他结构,可以把种子带到新的地方。

3. 水果被动物消化,但种子则经过消化道散落在其他地方。

4. 种子通过风吹动传播的植物,像蒲公英,往往会产生大量的种子,以确保种子可以吹到能令其发芽的地方。

5. 许多水生植物和生活在水边的植物结出可以漂浮并能被水携带的种子。

Unit 4

Lesson 1

Ⅰ. image self-employed match enterprises agriculture

Ⅱ. 1. A 2. D 3. A 4. D 5. A

Ⅲ. 现代食品是一项奇迹。你可以吃到来自世界各地的水果,新鲜得就如刚刚摘下来一样。农业和农民正在迅速改变。农业曾经一度是穷人的职业,现在已成为大产业,小型农场正逐渐消失,被农业企业所代替。消费者应该了解这种现代耕作方式的转变并且要知道它如何影响你。

Lesson 2

Ⅰ. hydrating yogurt cucumber summertime Chilled refreshing

Ⅱ. 1. Omitted.

2. It's high in water, very few calories, and rich of folate and vitamins A, C, and it neutralizes stomach acid.

3. These unassuming florets are packed with vitamins and phytonutrients that have been

shown to help lower cholesterol and fight cancer, including breast cancer.

Ⅲ. 1. 根据过去的经验,你应该每天喝八杯水。
   2. 喝大量的水很重要,尤其是在夏天。
   3. 树莓和蓝莓含水量都徘徊在85%,而黑莓略高于它们,大约为88.2%。
   4. 菠菜中含有丰富的叶黄素、纤维、钾以及促进大脑的叶酸,仅仅一杯生菠菜叶就含有每日摄入的维生素E(一种抵抗有害自由基分子的重要的抗氧化剂)的15%。

Lesson 3

Ⅰ. ensure   freshness   purity   packaged   guidelines

Ⅱ. 1. Tomato sauce is a kind of condiment served along with the food to enhance its taste and flavor.
   2. You would find purely tomato, onion, garlic, red chilly or soya bean flavor in tomato sauce.
   3. The basic ingredients are olive oil, peeled garlic, tomatoes and salt.

Ⅲ. 1. 由于它浓郁的味道和诱人的酸甜味,对于大多数菜肴来说,成了必不可少的一味调料,尤其是在家烹饪意大利面的时候。
   2. 番茄酱历来都是孩子们的最爱。成年人也无法抗拒番茄酱酸甜的风味。
   3. 你可以使用自制的番茄酱,也可以使用现成的。
   4. 大多数品牌公司和中小企业在产品上市之前都会对产品做测试。
   5. 在市场上推出番茄酱的同时,企业需要注意卫生和包装。

Lesson 4

Ⅰ. 1. celery   2. fiber   3. stalk   4. collagen   5. tryptophan   6. antioxidant   7. stalk   8. folate   9. micromutrient   10. calcium

Ⅱ. 1. 关节炎   2. 小病   3. 医学的   4. 医师   5. 肝脏   6. 发炎   7. 气喘   8. 症状   9. 平喘的   10. 血管

Ⅲ. 1. At first glance celery may seem rather unimpressive, but the more you look into its background and medicinal uses, the more you realize that we must have been misinformed on the usefulness of this plant.
   2. Our common celery stalk is mostly composed of about 83% water and a healthy amount of fiber.
   3. As our methods of researching whole foods develop, we are finding out more and more about what we are eating for dinner, and as we have just found out, celery is truly much more than just water and fiber!
   4. Celery is also a good source of vitamin C, and along with that comes all the benefits that vitamin C carries with it.

Ⅳ. 饮用生乳及生乳制品的消费安全问题几十年来一直存在。生乳消费的支持者和反对者都一直在提出理由证明他们的观点,双方都有充分的理由。

那些总体与农业,特别是与乳品业有着千丝万缕联系的人们懂得生乳消费的危险性。这根本就不是秘密,如果鲜奶没有经过合适的加工,会驻有危险的病原体,这些病原体会迅速严重地伤害我们的身体。尽管这样,还是有许多人继续要求消费生乳。具有讽刺意味的是,"买方负担风险"这一重大观念直到灾难临头才被我们意识到。

尽管存在风险,仍有许多人认为鲜乳及生乳制品的消费代表着饮用者健康饮食的基本权利和消费权利。他们的生活方式及价值观包括食品消费应该是"天然的",应该脱离实际存在的或有所察觉的潜在的掺假现象——那些来源于现代商业化农业的污染或折中。虽然牛奶加热杀菌法在绝大多数乳品业中已经是公众接受的常规,还是有相当比例的人口愿意接受存在于生乳消费中潜在的风险。

Ⅴ. Omitted.

Unit 5

Lesson 1

Ⅰ. 1. resources 2. confronted 3. distribute 4. process 5. control 6. manages 7. challenges 8. innovation 9. benefit 10. agriculture

Ⅱ. 他们的展示里,专家们有意选择了农业领域:在一个农业的立体模型上,一台带有工具的微型拖拉机在一块地上移动。农田边缘有两台平板电脑。展会与会者可以用它们来启动农场设备的自动化控制。农场模型上方悬挂着六个屏幕。他们显示自动化的过程,展示软件如何管理各项功能。今天使用的拖拉机和工具以广泛应用电子设备和软件为特色——这些被称为"嵌入式系统"。展会的座右铭是"看:软件工程的解释"。可视化可以帮助游客了解互连嵌入式系统和IT系统的挑战和解决方案。

Ⅲ. Omitted.

Lesson 2

Ⅰ. 1. lack 2. drive 3. efficiency 4. available 5. intelligent 6. decisions 7. commence 8. harvesting 9. identify 10. ultimate

Ⅱ. "我们开发的设备可以自主收集、分析和呈现这些信息,所以农民的主要工作都可以自动完成。"

新的一年,该团队将开始项目的第二阶段,包括应用本技术标准化操作农用拖拉机,以及感知环境、识别任何需要的操作,它们也能够执行这些操作本身,如施肥和喷洒农药、浇水、清扫和除草。

Unit 6

Lesson 1

Ⅰ. E D B C G F A

Ⅱ. 1. Agricultural mechanization enormously increased farm efficiency and productivity.

2. A farmer's day was labor-intensive, beginning well before sunrise and ending at sunset.

3. A large number of fatal injuries from tractors tipping over led to the design of rollover bars.

4. Engineering design for planting and harvesting was hampered by the wide variety of crops, all with different shapes and consistencies(e. g. , corn, soybeans, wheat, cotton, and tomatoes).

Lesson 2

Ⅰ. 1. organic farming 2. biodiversity 3. pesticide 4. organic manure 5. innovation 6. rotation 7. sustainable 8. dairy industry 9. vegetable 10. natural resources

Ⅱ. 1. 玉米 2. 黄豆 3. 烟雾 4. 长柄镰刀 5. (尤指不寻常的人和事)出现，到来 6. 内燃机 7. 秸秆压捆机 8. 谷物 9. 戏剧性地 10. 产量

Ⅲ. 1. intensive 2. comparison with 3. alternatives 4. approach 5. strategy 6. proved 7. polluted 8. income 9. sustainable 10. benefits

Ⅳ. 1. Organic farming is work intensive, in comparison with mechanical agriculture. Organic farming is best known for using natural alternatives to pesticides, such as natural predators and rotating crops. Organic farming is a type of farming relying on such techniques like plants rotation, natural plant foods and biological pest control.

2. It aims at the key of high-yield, high-returning, safety, sustainable farming of the future.

3. Organic food is more labor intensive since the farmers do not use pesticides, chemical fertilizers or drugs. Organic farming is the practice of producing food without the use of man made pesticides, herbicides, and fertilizers. Organic farming is an environmentally responsible approach to producing high-quality food and fiber.

4. Organic farming is a type of farming relying on such techniques like plants rotation, natural plant foods and biological pest control.

Ⅴ. Omitted.

Lesson 3

Ⅰ. 1. efficient 2. improve 3. damage 4. instead of 5. profit 6. out of control
7. balance 8. common 9. conservation 10. reasons

Ⅱ. 在欠发达、较贫穷的国家,情况截然相反:没有足够的粮食来养活所有人口。饥荒、饥饿和营养不良在这些国家很常见。有时候,其他国家会给这些国家提供紧急粮食供应,但这并不是长久之计。这些国家的农民需要生产自己的食物,采取能使他们的农场更高产的保护土地的方法。有时因为经济原因导致粮食短缺,但在其他情况下,仅仅是因为农民缺乏正确的教育。

Ⅲ. Organic farming can be defined as an approach to agriculture where the aim is to create integrated, humane, environmentally and economically sustainable agricultural production systems. Organic farming is a form of agriculture that relies on crop rotation, green manure, compost, biological pest control, and mechanical cultivation to maintain soil productivity and control pests.

Crop rotation is to grow specific groups of vegetables on a different piece of land each year. Groups are moved around in sequence, so they don't return to the same spot for at least three years. Crops are rotated to prevent exhausting soil nutrients while adding certain matters necessary for the following crops. Some plants have so few soil dwelling pests or disease that they serve as effective pest management tools in the rotation.

Unit 7

Lesson 1

Ⅰ. 1. The confusing part is that each variety needs a particular combination. For instance, a variety that needs many hours of summer light will not perform well in an area that receives fewer hours of light. Onion growers categorize onions in one of three ways: Short Day, Intermediate Day, and Long Day.

2. Onion plants are hardy and can withstand temperatures as low as 20 ℉. They should be set out 4 to 6 weeks prior to the date of the last average spring freeze. When you obtain onion plants, they should be dry. Onions are best grown on raised beds at least 4 inches high and 20 inches wide. Onions need a very fertile and well-balanced soil.

3. To reduce tearing when peeling or slicing an onion, chill for 30 minutes or cut off the top, but leave the root on.

4. Prolonged cooking takes the flavor out of onions. Cook only until they're tender when tested with a fork.

5. Onions are high in energy and water content. They are low in calories, and have a generous amount of B6, B1, and Folic acid. Onions contain chemicals which help fight the free radicals in our bodies. Free radicals cause disease and destruction to cells which are linked to at least 60 diseases. When a person eats at least 1/2 a raw onion a day, their HDL cholesterol goes up an average of 30%. Onions increase circulation, lower blood pressure, and prevent blood clotting.

Ⅱ. J D F I H A G B E C

Ⅲ. Omitted.

Lesson 2

Ⅰ. 1. 沙质土 2. 排水好的 3. 微酸性的 4. 肥沃的土壤 5. 堆肥 6. 白化 7. 相对湿度 8. 稻草 9. 灌溉 10. 杂草 11. 鳞茎 12. 质地 13. 暗处 14. 通风良好

Ⅱ. 1. 通过翻动堆肥或在堆肥上耕作来备好大蒜的苗床(一定要使用被彻底有氧分解,含有动物粪便及植物残体的堆肥,而不是杉木或红木)。

2. 以 18 英寸的间隔将大蒜(和象蒜)种成行。表皮鳞茎每行间隔 4 到 6 英寸。

3. 记住,大蒜要多浇水和施肥,但必须有良好的排水,否则它会腐烂。春季在大蒜开始长大时给大蒜施腐熟的有机肥或营养均衡的化肥。

4. 夏季当大蒜的叶子开始变黄的时候,两周不要浇水,然后把它拔下来。立刻将大蒜放置在阴凉处。

5. 正确的储存方法是将它编起来或者编织在一起(如果蒜瓣较软的话),然后把它悬挂在通风较好的阴凉处。

Lesson 3

Ⅰ. 黄瓜一般比西红柿生长更迅速,生产更早。它们还需要更高的温度,这意味着它们通常被作为春季或初夏作物。白天的温度应该在 80~85 ℉(夜间 65~75 ℉)。土壤温度最低应为 65 ℉。较低的温度将延缓植物生长和果实的发育。

植物最好在单独的容器里种植。由于种子通常非常昂贵,每个容器(1/4~1/2 英寸深)里放一粒种子。加水,盖上透明的聚乙烯,放在阴凉处。在 80~85 ℉下两到三天植物就长出来了。植物长出后去除覆盖的塑料,放在充足阳光下。

植物形成至少两片真正的叶子以后,把它们移植到苗床的固定位置。每株黄瓜间需要 6~8 平方英尺的空间。作物一般间隔 2 英尺,行间距 3 到 4 英尺。

Ⅱ. 1. concept 2. providing 3. efficiency 4. decreased 5. solution 6. pesticides 7. supply 8. Despite 9. remote 10. defined 11. traditional 12. nutrient solution 13. types 14. technique 15. method

Ⅲ. Omitted.

Unit 8

Lesson 1

Ⅰ. Bacteria, microscope, organic matter, energy, carbon dioxide, nitrogen

Ⅱ. 1. who 2. But 3. a 4. hid 5. Another 6. it 7. saying 8. will be 9. strength 10. Under

Ⅲ. 1. 在每盎司肥沃的土壤中有成千上万个小生物体。

2. 土壤中的微生物活动是相当复杂的，目前为止，我们还无法完全理解，但是我们确实知道它们可改善土壤的生产力。

3. 土壤藻类是含有叶绿素的非常小而简单的一种生物体，通过从空气中获取二氧化碳、从土壤中获取氮气来生长。

4. 某些类型的细菌能将空气中的氮转化为植物可用的氮化合物。

Lesson 2

Ⅰ. parasite, exist, present, starved, susceptible, wheat

Ⅱ. 1. was completed; is equipped 2. is; is 3. where; which 4. be; will be 5. are an 6. have; themselves 7. none; neither 8. acoustics, pajamas, measles; police, nucleus, cattle

Ⅲ. 1. 概括地说，植病的化学防治就是使用杀菌剂。杀菌剂可用于种子、植物或土壤。

2. 有些昆虫是造成严重植物病害的寄生虫载体，如控制甜菜中的蚜虫会减少黄化病的发病率。

3. 种子必须无病。尤其是小麦和大麦，因为它们能携带造成散黑穗病的霉菌，这种病往往根深蒂固，所以只能使用无病种子。

4. 在采取控制措施之前，先要了解造成疾病的原因。尽可能查明原因后，便可采用适当的预防或防治措施。

5. 作物轮作有助于避免寄生虫的积累。

Lesson 3

Ⅰ. crop yields, compete, harvesting, aggressive, spread, eradicating, plow up, cultivated, ploughing

Ⅱ. 1. have depended 2. is entitled to 3. to offer 4. arguing 5. the fatter 6. will have worked 7. be settled 8. did not poke

Ⅲ. 1. 土壤覆盖聚乙烯膜，然后注入挥发性化学药品。约24小时后把膜去掉，让土壤充分暴露在空气中数天再使用。

2. 微生物的复制和传播主要靠孢子、真菌和细菌,风、水、病株、粉屑、块茎、动物、人和昆虫都是一些疾病的传播媒介。
3. 在作物生产中控制杂草、病虫害是获得高产量必不可少的。
4. 杂草降低粮食产量是因为它们与农作物竞争水、土壤养分和光。
5. 耕种土壤的主要原因是除掉杂草。也可以通过除草剂这种化学药品来除草。除草剂有两种基本类型: 选择性和非选择性。

Unit 9

Lesson 1

Ⅰ. urban involved local offers elementary volunteer

Ⅱ. 1. reported 2. richest/wealthiest 3. by 4. when 5. by 6. However 7. younger 8. Both 9. acting 10. the

Ⅲ. 1. 都市农业是人们利用周围的空间创造的农场和后花园。
2. 什么是都市农业,这很容易定义,但大多数人想知道它具体是怎么操作的。
3. 很多时候,在经济不确定时期,花园都会被废弃,唯一要面对的问题是当花园又能赚钱时,很多业主会回来索要他们的花园。
4. 每一个都市花园都不一样,而且需求也不一样。
5. 现在许多地方都会把花园放在屋顶上或道路中间。

Lesson 2

Ⅰ. fresh environment transportation awarded issued Due to flavor

Ⅱ. 1. C 2. C 3. A 4. A 5. B 6. A 7. C 8. C 9. D 10. A

Ⅲ. 1. 南口农场,由北京首都农业集团管理,位于昌平南口镇的南部。这是农业部认定的第一个国家级绿色食品生产示范基地,也是北京郊区最大的国有农场。
2. "燕广"苹果由于其多汁、肉脆、酸甜爽口、易储存的特点在北京乃至全国都很受欢迎。
3. 与许多政府机构、研究机构、大学和企业建立了长期的合作关系,不遗余力地提供花卉鉴赏、团队建设活动、科普教育、餐饮、旅游和农家乐。

Lesson 3

Ⅰ. covers kilometers core involves agriculture integrates township

Ⅱ. 1—5 ACBDC 6—10 DCBAC

Ⅲ. 1. 小汤山现代农业科技示范园建于 1998 年,在 2001 年被包括中国科学技术部在内的六部委正式命名为昌平国家生态农业园。

2. 园区在小汤山，靠近华北平原的北部。

3. 园区定位为"展示现代农业科技，促进周边地区发展，促进农业旅游"。

4. 它是一个集现代农业技术、优化农业产业、增加农民收入的基地。

5. 该区一年四季对游客开放。游客可以从中了解到最新最先进的农业科学技术并且可以参观大型温室，里面有各种各样的名贵花卉、蔬菜、水果，并有机会采摘水果、体验现代农业生活。

Unit 10

Lesson 1

Ⅰ. Resume

Name: Wang Xiaohua

Sex: Male

Date of Birth: June 15, 1998

Address: 56 Binhai Road, Dongfang City

Mobile phone: 18699112300

E-mail: wangxiaohua@ 163. com

Education Background

September 2010 to July 2013, Majoring in Seeds Producing & Marketing in Dongfang Vocational College

September 2006 to July 2009, No. 1 High School in Dongfang City

2012, passing computer exam and getting computer certificate

Awards

2010 to 2011, winning yearly scholarships

Work Experience

January 2012 to April 2012, internship in Dunhuang Seed Industry Co., Ltd.

Professional Interests

Having intensive interest in traveling and reading

Ⅱ. Omitted.

Lesson 2

Ⅰ. Dear Sir,

I've learned from *China Daily* on September 20th that your company is employing foreign seed production technician and foreign trade sales manager. We all know that Dunhuang Seed Industry Co. , Ltd. is a well-known seed production enterprise with good reputation. I am writing to apply for the positions' interview.

As requested, I am enclosing a completed job application, my certification and my resume.

I graduated with crop production technology from Jiuquan Vocational and Technical College. During my college, I passed the practical English test and gained band A certificate. I've gotten excellent result in English oral contest. Also, I am skilled in computer and have proficiency in the use of WPS, Excel, Photoshop etc. As a very open and forthcoming student, I served as president of the student union with strong ability to work.

I hope you will give me an opportunity to meet you and attend a personal interview and I will work hard and turn an asset to your company.

Yours faithfully,

Zhang Liang

Ⅱ.

No. 450 Dongfang Road,
Haidong City
June 6th, 2019

Dear Sir,

I am writing to apply for the sales manager position. As requested, I am enclosing a completed job application, my certification, and my resume.

The opportunity presented in this listing is very interesting, and I believe that my strong technical experience and education will make me a very competitive candidate for this position. The key strengths that I possess for success in this position include:

- I graduated from Dongfang College in 2017.
- Now I am working as a mechanical engineer in SFG and responsible for maintaining the machine and equipment.
- I was trained with technology in both domestic and foreign countries during my working.
- I'm good at communicating with others and cooperating with team members.

Please see my resume for additional information on my experience.

I can be reached anytime via email at xialei000@163.com or my cell phone, 13670080000. I can start my new job on October 15th, 2017.

I will appreciate a lot if you can give me an opportunity to attend the interview. Thank you for your time and consideration. I look forward to your reply.

Yours faithfully,

Wang Jun

Lesson 3

Ⅰ.

We have pleasure in receiving your order of Spet. 15 for tomato juice and you are welcome to be one of our clients.

We confirm having supplied you the above-mentioned goods at the price listed in your letter and have effected shipment by mv. "Princess" next week. We firmly believe you will be utterly satisfied when you received them.

You may don't know the scope of our business activities. Now we send you, under cover, a copy of catalog. We hope that this initial order will result in more business connections and pleasant working relations.

Ⅱ.

June 19, 2019

Dear Mr. John Brown,

First of all, I'd like to show my great thanks for your order to our latest products. All the goods you've ordered have been shipped in good condition and will be delivered after one week. We expect your reply when receiving the goods.

We hope that we can cooperate in more fields and looking forward to receiving your continuous order in the near future.

Sincerely yours,

Wang Ming

Manager of Sales Department

参考文献

[1] Julie Kerr Casper. Agriculture：The Food We Grow and Animals We Raise[M]. NewYork：Chelsea House Publishers, 2007.

[2] J. A. R. Lockhart, A. J. L. Wiseman. Introduction to Crop Husbandry [M]. Robert Maxwell：Pergamon Press, 1978.

[3] 兰天. 外贸英语函电[M]. 8 版.大连：东北财经大学出版社,2018.

[4] 教育部《农林英语》教材编写组. 农林英语[M]. 北京：高等教育出版社,2001.

[5] 张永萍,吴江梅. 农林英语[M]. 北京：北京大学出版社,2009.

[6] 夏家驷. 农业专业英语[M]. 武汉：武汉大学出版社,2011.